# SOLAR HOME HEATING Basics

DAN CHIRAS

Illustrations by Anil Rao, Ph.D.

Cover design by Diane McIntosh.

Inset image: H. Mark Weidman Photography/Alamy; Background image: (leaf) © iStock (alpha cat); Illustration: © iStock (Diane Labombarbe)

Printed in Canada. First printing March 2012.

Paperback ISBN: 978-0-86571-663-6
eISBN: 978-1-55092-508-1

Inquiries regarding requests to reprint all or part of *Solar Home Heating Basics* should be addressed to New Society Publishers at the address below.

To order directly from the publishers, please call toll-free (North America) 1-800-567-6772, or order online at www.newsociety.com

Any other inquiries can be directed by mail to:

New Society Publishers
P.O. Box 189, Gabriola Island, BC V0R 1X0, Canada
(250) 247-9737

Library and Archives Canada Cataloguing in Publication

Chiras, Daniel D.
Solar home heating basics / Dan Chiras ; illustrations by Anil Rao.

Includes bibliographical references and index.
ISBN 978-0-86571-663-6

1. Solar houses. 2. Solar heating. I. Title.

TH7414.C45 2012 697'.78 C2011-908623-9

New Society Publishers' mission is to publish books that contribute in fundamental ways to building an ecologically sustainable and just society, and to do so with the least possible impact on the environment, in a manner that models this vision. We are committed to doing this not just through education, but through action. The interior pages of our bound books are printed on Forest Stewardship Council®-registered acid-free paper that is **100% post-consumer recycled** (100% old growth forest-free), processed chlorine free, and printed with vegetable-based, low-VOC inks, with covers produced using FSC®-registered stock. New Society also works to reduce its carbon footprint, and purchases carbon offsets based on an annual audit to ensure a carbon neutral footprint. For further information, or to browse our full list of books and purchase securely, visit our website at: www.newsociety.com

# Contents

CHAPTER 1

# How Will We Heat Our Homes?

Let's face it: times are changing. And, by most measures, they're not changing for the better. Ever-increasing levels of carbon dioxide are driving Earth's climate into a frenzy of pernicious and costly changes. Ever-more-violent hurricanes and a steep increase in the number of devastating tornados and floods, as well as droughts, crop failures, raging forest fires, spreading deserts, and the expansion of insect-borne diseases — all due to human-fostered climate change — are costing humankind dearly in human lives and dollars.

Meanwhile, energy demand is skyrocketing. Because much of the world's energy currently derives from burning fossil fuels, costly climate disruptions are bound to escalate. The price tag of our dependence — some say our addiction — to fossil fuels is bound to reach epic proportions, unless we do something dramatic, and very soon.

In addition to the serious problems created by climate changes, the United States and other countries are quickly depleting their limited reserves of natural gas used for heating our homes and water. Although ads that run on US television stations promise an abundance of natural gas associated with domestic shale reserves, the ads fail to note that these huge deposits of natural gas are located in extremely deep shale deposits in the eastern United States and that extracting this resource will be extremely expensive, if not impossible.

And then there's the "turmoil" in the Middle East as the United States — and some of its staunch allies — attempt to secure energy resources in a region in which we've burned way too many bridges and where true friends are few and far between. The United States will likely continue to wage exceedingly expensive wars in the Middle East to secure the fuels required by our often recklessly wasteful energy-hungry society.

But all the news is not grim. There is a bit of good news that deserves attention: Slowly but surely, many nations are turning to clean, affordable, and renewable energy supplies — even the United States. The two darlings of renewable energy are solar electricity and wind energy (Figures 1-1a and b). Large commercial wind farms and solar arrays are popping up like daisies in a summer meadow. These new renewable energy sources provide enormous amounts of electricity to power our future.

Despite the monumental importance of the shift to wind and solar electricity, reliance on these clean renewable resources won't solve a fundamental need facing many nations, namely, a carbon-free means of providing heat to homes and other buildings — and not just new homes and office buildings but the millions upon millions of existing homes, condos, townhouses, apartments, copy shops, grocery stores, schools, businesses, and the like.

Fact of the matter is, few buildings are currently heated with electricity. Electricity is too costly. Natural gas is a much more efficient means of heating, so most buildings in the United States and Canada are currently heated with natural gas.

Some proponents of solar- and wind-powered electricity argue that we could use the electricity generated from large commercial solar systems and massive wind farms to heat our

DAN CHIRAS

Fig. 1-1a and 1-1b: *(a) This large PV array near Golden, Colorado and (b) turbines in a giant wind farm in central Kansas are vital to the new sustainable energy economy.*

homes and offices, but electric heating systems consume enormous quantities of electricity to get the job done, and are the least efficient means of heating buildings, bar none.

## The Purpose of this Book

Fortunately, far more cost-effective means of heating homes and offices are currently available. Some are remarkably inexpensive. That's what this book is about: finding clean, reliable, economical, and renewable solutions to heating the buildings in which we live, work, and play. In this book, I explore four extremely clean, highly affordable, and infinitely renewable solutions to our heating dilemma.

Like the other books in this series, *Solar Home Heating Basics* does not dwell on the menacing problems we face. Like virtually all of the books I have written, this book is about *solutions* — affordable solutions that make sense both now and in the long-term. It is written for those who are interested in ending the world's outrageously costly addiction to fossil fuels. It is written for those who want to dramatically reduce their carbon footprint and the many social, economic, and environmental impacts of global climate change. In a word, this book is for those who are interested in ensuring a sustainable future — one that's good for "all the children of all species for all time," to borrow a phrase from green architect Bill McDonough.

This book is not a prescription for a national or international energy plan. It's a prescription for a *personal* energy plan. It focuses solely on personal solutions — steps you and I can take today and tomorrow and the next day to reduce our fuel bills, create higher levels of comfort, and build an energy system that doesn't bankrupt humankind and destroy the life-support systems of the planet. These solutions can and should be incorporated into national energy plans, but I leave that to others with more patience for the slow and arduous pace of political reform.

This book is written both for readers who are building a new home or business or, as is more likely the case, those who want to retrofit an existing home or business to cut fuel bills, increase comfort levels, and contribute to a better society and a better world.

Although the book is called *Solar* Home *Heating Basics*, all the technologies described here could be applied to other types of buildings, including office buildings, shops, or schools.

## Organization of the Book

In Chapter 2, I explore energy efficiency and conservation and their importance for anyone contemplating a solar heating system — or any renewable energy system, for that matter. This chapter explains what energy is and the units of measurement used to describe it. As you shall see, energy-efficiency measures are far cheaper than renewable energy systems, and by offsetting our demand, they help reduce the initial cost of renewable energy systems required to power our homes.

In Chapter 3, I turn my attention to one of the most important things individuals can do to save energy and increase comfort levels: weatherization. It is a step few people take, but one that should be taken *before* installing a solar heating system. Weatherization means sealing up the many energy-wasting cracks and holes in the buildings we occupy. These energy thieves are as ubiquitous as mosquitoes in Minnesota in July — and once you know about them, they're just as annoying. Moreover, they are costing us a fortune each year by jacking up our heating and cooling costs. Sealing the leaks is a prerequisite to successful and cost-effective solar home heating. This chapter and the next outline many practical steps you can take to dramatically reduce energy consumption required to maintain comfort through many a long, cold winter.

In Chapter 4, I cover the second prerequisite to economical solar home heating: insulation. Few of us would think about spending more than a minute or two outdoors on a cold winter day without a heavy coat. But, in effect, most of our homes are standing out in the cold with nothing more on than a thin flannel shirt and a pair of shorts. Sparingly clad, they're forced to endure subfreezing weather day after day, year after year. The only way we can stay warm inside our homes is to bundle up or crank up the heat. In this chapter, I tell you where to apply insulation, what types of insulation are available, how to install insulation, and why insulation is just as important in hot climates as it is in cold ones. The chapter also includes information about the various incentives available to homeowners who add insulation to their homes.

In Chapter 5, I explore the source of free energy: the Sun. I discuss solar radiation and other important concepts that will provide you with the information you need to understand your solar home heating options.

In Chapter 6, I look at the simplest, least expensive, and most reliable solar heating technology — a technique known as *passive solar heating.* You'll learn what passive solar is, the basic principles of passive solar design, and how to incorporate passive solar into new buildings. I also give you the costs and benefits of this Earth-friendly approach, and some of the common mistakes made in passive solar design (and, perhaps even more importantly, ways to avoid them). You'll also learn why everything you do to passively heat a home will help cool it in the summer.

Chapter 7 covers ways to retrofit a building to incorporate passive solar heating. It's tricky, I'll warn you now, but achievable. (This chapter appeared previously in a slightly different form in one of my favorite magazines, *Home Power.*)

In Chapter 8, I present another affordable option for providing space heat, a technology known as *solar hot air heating.* These systems rely on *solar collectors.* (The term "collector" in this book refers to a variety of devices. What they have in common is that they are technologies used to capture the Sun's energy.) In this chapter, I describe the types of solar hot air collectors, how they operate, and how they are installed. I'll also discuss the limitations of this technology, and, as in other chapters, you'll learn about the costs and benefits of these systems.

Chapter 9 looks at the third main solar home heating solution, solar hot water systems — commonly referred to as *solar thermal systems*. In this chapter, I discuss the types of solar hot water systems, types of collectors, requirements for a successful installation, the costs and benefits, and a lot more.

Chapter 10 covers conventional heating systems. They're almost always required in solar-heated homes to provide backup heat during cloudy periods and to satisfy building code requirements. In this chapter, I discuss the most energy-efficient and cost-effective backup heating systems (including systems that rely on renewable resources such as biodiesel) and incentives for those who install energy-efficient boilers and furnaces.

At the end of the book, I've included a Resource Guide that presents numerous books, articles, and other resources you can consult to learn more.

CHAPTER 2

# A PRIMER ON ENERGY AND ENERGY EFFICIENCY

In my classes on solar and wind energy at The Evergreen Institute, I continually remind my students of the virtues of energy efficiency and conservation — measures that reduce household energy use. In fact, like other solar advocates, I consider reductions in energy demand as a prerequisite to renewable energy. Accordingly, I encourage my students to make their homes (or, in the case of future solar- or wind-system installers, their clients' homes) more energy efficient *before* they even think about installing a renewable energy system. I give the same advice to my readers, you included.

This chapter explores the rationale behind my push for efficiency first, advice I hope you will take before you retrofit your home with any of the solar home heating options discussed in this book. This chapter also presents key energy concepts and the terminology that you need to know when studying your options and making decisions on what is appropriate for your home or business.

## Energy Conservation and Energy Efficiency

Reductions in energy consumption can be achieved in two ways: by employing energy conservation measures and by employing efficiency measures.

In home heating, *energy conservation* refers to the simplest measures one can use to cut down on energy use — for example, turning down the thermostat in winter when you leave the house for work each morning. Energy conservation measures typically include changes in behavior, such as turning off lights or wearing a sweater, insulated underwear, and heavy socks on cold winter days. It might include something as simple as donning one of those quirky Snuggies while reading or watching TV, or switching on a ceiling fan to force hot air down from the ceiling to the occupant level. I group all similar efforts within the broad category of energy conservation.

While energy conservation often entails behavioral changes, energy *efficiency* relies principally on technology to trim demand. Energy-efficient technologies are those that provide services with the least amount of energy.

Energy efficiency is a measure of energy in to energy out — that is, how much energy a device uses to provide a service, such as lighting or heating, compared to the amount of service we receive. A 100-watt incandescent light bulb, for instance, requires 100 watts of electricity, but delivers only about 5 to 10 watts of light energy. So, it is only 5% to 10% efficient. The remainder of the energy consumed by the light bulb is emitted as heat.

The more efficient a device is, the greater the conversion of energy into some useful end product. Substituting an energy-efficient LED floodlight to illuminate your front steps and walkway, for instance, can reduce the wattage of an outdoor light from 150 to 10 watts (Figure 2-1). You still receive the light you need to safely navigate the toy-littered front steps at night, but you do so with a lot less energy. The LED light produces proportionally more light per unit of energy it consumes than the incandescent lights that have been in use for many years. As another example, an energy-efficient furnace delivers more heat per unit of the natural gas or propane it burns than

DAN CHIRAS

Fig. 2-1: *This light consumes 10 watts but produces as much light as a 150-watt conventional floodlight.*

its less-efficient counterpart. You stay warm with less fuel and save heaps of money in the process. So why spend time talking about efficiency in a book on solar home heating solutions?

The answer is simple. *It costs less to save energy than to buy energy — a lot less.* For example, it is much cheaper to don a comfortable wool sweater, or to wrap up in a blanket as you read or watch TV, than it is to crank up the thermostat four degrees. Yet they both have the same effect.

Study after study shows that it is much more economical to seal up the leaks in the walls, foundations, and roofs — known as the *building envelope* — with a few $5 tubes of caulk than it is to crank up the heat or even to generate heat from a renewable energy technology such as a solar hot water system.

## Short-Term and Long-Term Benefits of Efficiency and Conservation

Efforts to reduce energy consumption can save you a fortune — and reduce your carbon footprint. In the short term, energy efficiency and conservation measures can save money by substantially reducing the size of the renewable energy system you'd need to install to meet your needs. Such measures often save enormous amounts of money. So remember, slashing energy use dramatically reduces the size and initial cost of renewable energy systems.

The way it works is simple: by lowering household energy demand, efficiency and conservation measures reduce the amount of energy your renewable energy system will need to supply to maintain comfort in your home. That means you'll be able to install a smaller system to meet your needs. The smaller the system, the lower the initial cost. Investing a few thousand dollars in energy-efficiency measures to cut overall heating demand by 25% could reduce the cost of that $20,000 solar hot water heating system you were contemplating by as much as $5,000.

Measures to reduce energy consumption save in the long-term as well — by reducing your monthly fuel bills. If, for instance, you install a solar hot water system, you'll still need to use your furnace once in a while — say, during long, cold, cloudy periods when there's just not enough sunlight to provide space heat. The more efficient your home is, however, the shorter the run time, and the less fuel you'll need. The less fuel you need for backup heat, the more you'll save.

If you choose *not* to pursue a solar home heating option, efficiency and conservation will by themselves save you a lot of money over the long haul. The weather stripping you install around leaky double-hung windows or leaky doors, the caulk you apply at the base of your walls, and the host of other measures discussed in the next two chapters can save you huge amounts of money and make your home much more energy efficient — reducing the waste of expensive energy. So take these steps, at the very least.

## *Efficiency isn't Sexy, but it Really Works*

Over the years, as I've been extolling the virtues of energy conservation and efficiency, several students in my classes have responded that these measures just "aren't as sexy as renewable energy." It's hard to deny. A brand new solar hot water system or PV module is an exciting addition to a home and will gain a lot of attention — even accolades. Not many people will remark favorably (or at all) on your new caulk and weather stripping.

Without a doubt, modern humans are enamored of shiny gadgets, which are a lot sexier than a bead of caulk or some lowly weather stripping applied around doors and windows to seal up heat-robbing air leaks. However, all the sex appeal of a renewable energy system quickly fades when you compare the price of achieving the same energy reductions with efficiency measures. I learned this lesson many years ago after moving into a passive solar home in Colorado. Although the house was heated by the Sun and was pretty efficient, it was all-electric. It had an electric water heater, an electric stove, and electric baseboard heaters for backup space heat. The previous owner had not installed an energy-efficient refrigerator or any energy-efficient lighting. To power the home with solar electricity, I would have needed a $50,000 system. A couple of years later, I built a new home of similar size, but this one was designed and constructed to be as energy efficient as humanly possible. To power that home, I was able to meet my demands for electricity with a $16,000 solar electric system, all thanks to careful attention to energy conservation and energy efficiency and just a few thousand dollars additional investment. I got my solar system at a much lower initial cost thanks to the more affordable efficiency measures that trimmed demand.

Conservation and efficiency may not be sexy, but they work really well, and they save lots of money. Besides reducing initial costs, these two measures continue to save me money every year. From 1996 until 2010, for example, I saved approximately $20,000 on heating and cooling costs for my home in Evergreen, Colorado in the foothills of the Rocky Mountains. My savings have paid for the slightly higher cost of building an airtight, superinsulated, earth-sheltered passive solar home ($1,000 to 2,000) and the solar electric system ($16,000), which generated about $4,000 worth of electricity during that period.

Energy conservation and energy efficiency offer the best of both worlds. They save money initially by reducing the cost of renewable energy systems, and they save a ton of money in the long-term. It couldn't get much better.

### *Energy Efficiency and Conservation are Renewable Resources*

Energy efficiency and conservation are renewable "sources" of energy, too. That new energy-efficient compact fluorescent light bulb or that new Energy Star refrigerator you install, for example, will continue to save energy year after year after year. And remember, every bit of energy you save through such efforts — now and in the future — is freed up for others to use. Saved energy is energy that doesn't need to be extracted from the Earth's declining supply of coal or natural gas.

If that doesn't make sense, think of it this way: Imagine that a small city with 50,000 homes expands by 100 new homes a year. Suppose the homes are heated by natural gas. To heat these new homes, additional natural gas deposits must be located, and natural gas must be extracted and piped to the new homes. Suppose, however, that the citizens of our hypothetical city embark on a citywide energy-efficiency and conservation effort. They enact various measures that reduce their natural gas demand by the same amount required to heat the 100 new homes. The net result is that the new homes are supplied by efficiency measures. And, there's no need to extract more natural gas.

Conservation and efficiency are, in effect, the same as developing new energy resources. Annual savings go on year after year after year, too, creating an essentially renewable supply of energy. What is more, this strategy doesn't increase the city's carbon footprint. Per capita energy use and per capita carbon emissions actually decline.

## Understanding Energy

Before you can save energy, it is important to first know what energy is and what the terms are that are used to describe it — specifically, the units of measurement.

### *What is Energy?*

In simplest terms, energy is the "stuff" that allows us to do work. It helps us heat and cool our homes. It powers our lights and appliances. It powers electronic devices like cell phones, computers, fax machines, and stereos. It powers motors, busses, and even electric cars.

Most of the energy we use comes from the Sun. Coal, natural gas, and oil are nothing more than a mixture of organic materials that contain ancient solar energy. How so?

Astonishingly, the energy released when fossil fuels are burned is ancient solar energy. That's because fossil fuels are made from plant matter (in the case of coal and natural gas) and photosynthetic algae (in the case of oil). Plants and algae get their energy from the Sun. During photosynthesis, this energy is used to make organic molecules that make up the bodies of plants and algae. When these organisms die, they are covered by sediment. Over time, heat and pressure convert the organic matter into fossil fuels. When we burn these fuels, we release the ancient solar energy.

DAN CHIRAS

Fig. 2-2: *The south-facing windows in this passive solar home in the Northern Hemisphere (Carbondale, Colorado) allow the low-angled winter sun to enter, providing heat during the cold winter months.*

Unfortunately, utilizing this form of solar energy also releases billions of tons of carbon dioxide and other pollutants. They, in turn, have made quite a mess of the Earth's atmosphere, the air we and all the rest of Earth's creatures breathe. Carbon dioxide, of course, is also responsible for mucking up the planet's climate. As most readers already know, carbon dioxide is a greenhouse gas that traps heat in the Earth's oceans and atmosphere. This extra heat wreaks havoc on the climate.

We can, however, tap *directly* into solar energy without all the garbage carbon. That's what this book is all about: finding ways to heat our homes from carbon free sunlight beaming down from the sky (Figure 2-2).

## *Putting Energy to Good Use*

Energy comes in many different forms. There's sunlight energy, electrical energy, heat, light, and even coal, oil, and natural gas. These last three are considered *raw energy.* All forms of energy have one thing in common: they permit us to do work. However, most forms of raw energy are pretty useless to humankind. A lump of coal or a gallon of gas, by itself, is hardly worth the money we pay for it.

Raw forms of energy become useful (allow us to perform work) when we convert them into other, more useful forms. When coal is burned in a power plant to produce heat, for instance, the heat is used to boil water. Steam from the superheated water is used to drive a turbine that is attached to a generator that then produces electricity. Sunlight by itself isn't that useful either. However, Sun streaming into the south-facing windows of a home is converted into heat that can warm our living spaces, making life more bearable. Sunlight bombarding a solar electric module, on the other hand, causes electrons to move, creating an electrical current that can power lights, appliances, electronics, and even electric cars. All this is to say that it is the *conversion* of raw energy into useful energy that matters to us the most.

Over the years, humans have devised many amazingly ingenious technologies to convert "raw" solar energy into useful forms. This book is about three of them: passive solar homes, solar hot air systems, and solar thermal systems (also called "solar hot water" systems). Each system requires a device to collect and then convert solar energy into a more useful form of energy — heat.

One thing to remember, though, is that solar energy is a *low density* form of energy. That is, it doesn't contain a lot of energy per unit volume compared to coal or oil or even natural gas. Another fact to keep in mind when working with energy is that each time we convert one form of energy to another, we lose some energy — often, a lot of energy. Put another way, no energy conversion is 100% efficient. Far from it. Your car, for instance, only converts about 20% to 30% of the energy it consumes into forward motion. Diesel engines convert a bit more (about 40%) of the energy they consume into forward motion. During the conversion of these fuels into mechanical energy, there's obviously a huge amount lost. Energy is lost as heat, which simply radiates out into outer space.

Because of the diffuse nature of solar energy and the loss of energy during conversions, to make the most of solar energy, we must do everything within our power to retain the useful energy we obtain from the Sun and to use it efficiently.

As you will see, all of the solar home heating technologies discussed in this book are designed to gather up diffuse solar energy and then convert that energy into a more directly useful form, specifically, heat. Like all other technologies, solar heating equipment doesn't convert 100% of the solar energy it captures into useful energy. But if we are smart about it, we can keep our homes warm and cozy in the winter, all thanks to conservation, efficiency, and the Sun.

To make the most of the heat we gain in these systems, we need to be sure that we retain that heat where it belongs, so we can enjoy its benefits. To do that, we need to seal up our homes and offices as tightly as possible while maintaining a fresh air supply. We also need to insulate the daylights out of our buildings. My mantra is this: Insulate, Insulate, Insulate!

Weatherization, air sealing and insulation, are keys to making solar heating systems work.

Period.

Attempting to use solar heating without weatherization is like trying to fill a leaky bucket with water.

### *Units of Measurement*

Now that you understand a little about energy and the importance of energy efficiency and energy conservation, let's take a look at some common units of measurement. You'll very likely encounter these units in discussions with energy auditors or installers.

In the heating and cooling arena, the most important unit of measurement is the British Thermal Unit, or BTU. A BTU is the amount of energy it takes to raise the temperature of one gallon of water one degree Fahrenheit (specifically from 59° to 60° at one atmosphere of pressure).

BTUs are used to measure the heat content of fuel. Each type of fuel contains a certain number of BTUs per gallon or cubic foot or pound. A gallon of gasoline, for instance, releases 115,000 BTUs when it is burned; a gallon of diesel fuel contains 130,500 BTUs. A cord of firewood cut from a pine tree contains 17 million BTUs of heat that can be liberated during combustion. A hardwood tree like an oak would contain nearly twice as much energy.

BTUs are also used to describe the heat output of appliances such as water heaters or furnaces. A 60,000-BTU heater, for instance, produces 60,000 BTUs per hour when running at full speed. Obviously, the higher a furnace's or boiler's BTU rating, the more heat it will produce.

If your home is heated with natural gas, you'll note that your fuel charge is based not on BTUs, but on another unit of measurement, the *therm*. A therm is 100,000 BTUs.

In this book, I'll primarily be concerned with BTUs, since my focus is on home heating. I should mention, however, that BTUs are also used to measure heat *extracted* from a building. This is known as *cooling capacity*. A 10,000-BTU air conditioner removes 10,000 BTUs of heat from a building per hour. The higher the BTU rating, the greater the cooling capacity.

Cooling capacity is also expressed in tons. This term refers to the cooling capacity of a ton of ice. An air conditioner with a one-ton cooling capacity, for example, removes 12,000 BTUs/hour. In other words, a one-ton air conditioning unit is the same as a 12,000-BTU unit.

Just as BTUs can be converted to therms, BTUs can also be converted to another common unit of measurement:

kilowatt-hours. To understand what a kilowatt-hour is, we can begin with a more familiar measurement: watts. All of us have purchased light bulbs and other electrical devices, such as microwave ovens and hair dryers, by their wattage. A 100-watt light bulb, for instance, is so rated because it consumes 100 watts of electricity continuously when turned on. A 1,000-watt microwave oven consumes 1,000 watts of electricity continuously, when operating. A 1,200-watt hair dryer, consumes 1,200 watts of electricity.

Watts can be confusing to newcomers to electricity. Bear in mind that it is not a measure of quantity, like pounds or liters. It is a *rate*. More specifically, it is the instantaneous rate of power consumption. It's a lot like the speed of a car. If you are traveling at 50 miles per hour on a highway, that is a rate of movement. It doesn't say a thing about how far you will travel. It just tells you at that instant, you are traveling at 50 miles per hour. If you kept that speed up for an hour, you would travel 50 miles.

Utility companies aren't concerned with instantaneous power consumption of your home, however. They're interested in how much power you consume *over time* — specifically, the monthly billing period. Power consumption over time is measured in kilowatt-hours. The following example illustrates what a kilowatt-hour is:

If a 100-watt light bulb runs for 1 hour, it consumes 100 watts. This is expressed as 100 watt-hours of electricity.

If that bulb is left running for 10 hours, it consumes 1,000 watt-hours (100 watts x 10 hours = 1,000 watt-hours). If that bulb is left running for 20 hours, it consumes 2,000 watt-hours.

Because a *kilo* is equal to one thousand, 1,000 watt-hours can also be referred to as a kilowatt-hour. A kilowatt-hour is abbreviated *kWh*.

Two thousand watt-hours is 2 kilowatt-hours or 2 kWh. Twelve thousand watt-hours is 12 kilowatt-hours or 12 kWh.

It is kilowatt-hours that utility meters track and what utility companies charge you for each month.

Kilowatt-hours is the unit of measurement typically used to assess electrical consumption. Interestingly, as noted above, kilowatt-hours can also be converted to BTUs and vice versa. We most often convert BTUs to kilowatt-hours. A kilowatt-hour of electricity is 3,412 BTUs.

## Conclusion

Energy conservation measures like sealing up the leaks are key to successful and economical solar heating. Energy conservation and efficiency are to solar what walking is to running. Ignore these important steps, and you're bound to fall. You'll have to oversize your system to create heat, the vast majority of which will just leak out into the bitter cold winter air.

CHAPTER 3

# The Prerequisite to Renewable Energy: Seal up the Leaks!

Before you even consider installing a solar heating system, take time to seal up the leaks in the building envelope. When that's done, the next step is to beef up the insulation.

The building envelope is the outside skin of a building. It includes the roof, walls, and foundation. It separates and protects the occupants of a building from the cold of winter, the blistering heat of summer, and all the rest of the unpleasantries that Mother Nature sends our way, including wind, rain, snow, and sleet. An airtight building envelope serves the same purpose as a windbreaker or the wind-protective fabric of a coat. It prevents wind from robbing us of valuable heat.

In this chapter, I'll explore this crucial first step in successful solar home heating — sealing up the leaks. I discuss how big the problem is, the many sources of leaks in buildings, and how to locate and seal them. You may be surprised to learn how little it costs and how much money it can save.

## How Leaky are Our Buildings?

Most buildings are like Swiss cheese — that is, they are riddled with holes. Some of these troublesome holes are nothing more than tiny cracks that run along the junction between walls and floors (Figure 3-1). Others are more substantial

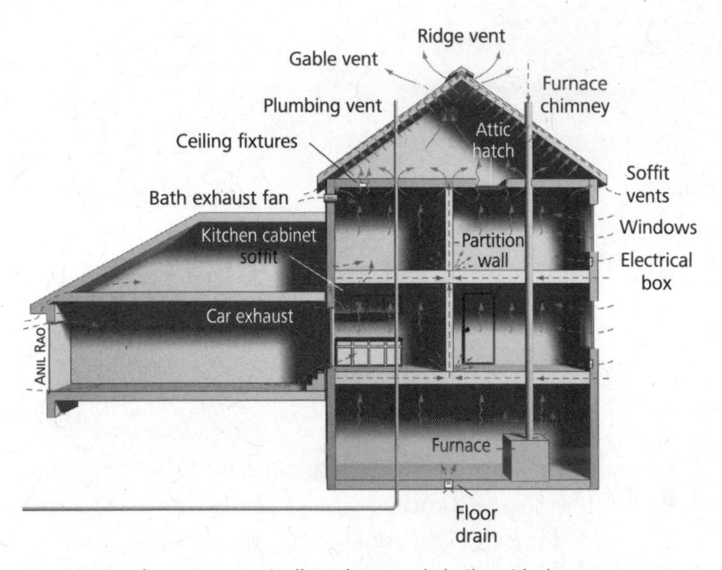


Fig. 3-1: *Our homes are typically rather poorly built, with dozens upon dozens of leaks in the building envelope. Sealing the leaks in both new and existing buildings can dramatically reduce energy consumption, lower fuel bills, and reduce the size of a solar heating system.*

gaps, like the ones that run the entire perimeter of exterior doors. As you examine your home, you'll probably find many openings. Some may be small holes in the building envelope, such as where pipes or electrical wires enter. Others, like missing panes of glass in basement windows, may be large enough for a raccoon or cat to enter.

When added up, the gaps in the building envelope can be quite substantial. To put things in perspective, remember that a one eighth-inch gap that runs 8 feet is equivalent to a one-square-foot hole in the building envelope. If you could add up all the cracks in the building envelope of older homes — many of those built in the 1800s and 1900s — they'd be equal to a 3-foot x 3-foot window that remains open 24 hours a day, 365 days a year. If your home was built before the 1980s, chances are it is riddled with holes that are robbing you blind and making your home a lot more uncomfortable — not just in the winter, but in the dead of summer, too.

While "modern" builders have paid more attention to the airtightness, the buildings they put up are still pretty leaky. If the leaks in the building envelopes of many modern homes were added up, they'd be equal to a 2-foot x 2-foot window open 24 hours a day, 7 days a week, and 365 days a year!

No one in their right mind would leave a 3-foot x 3-foot — or 2-foot x 2-foot — window open on a cold winter day. It would be sheer lunacy. But we are effectively doing just that. By not sealing up the dozens of leaks in our homes, we're effectively leaving large, energy-wasting windows open all year.

In the winter, when the wind blows, cold air enters through the Swiss cheese exteriors of our homes and offices. At other times, warm air generated by our furnaces and boilers leaks out through these gaps. In the summer, leaks in a building envelope allow hot air to enter on windy days. On calm windless days, leaks allow cool, air-conditioned air to stream out. We compensate by simply turning up the heat or turning down the air conditioner, putting up with drafts and discomfort, and grimacing when we open the utility bill.

If energy were free, and global warming and the global climate crisis it is spawning didn't exist, this waste might be tolerable. Faced with ever-rising fuel prices and the ever-worsening climate crisis, this waste is foolish and reckless. All the tiny leaks, when added to those in the millions of similar homes and businesses throughout the world, has a huge impact on annual energy consumption and thus an enormous impact on our planet, our lives, our economy, and our long-term future.

## The Economic Savings of Airsealing

Most people I've talked with over the years seem blissfully unaware, or at least highly tolerant, of the multiple — and costly — leaks in the envelope of the buildings they work and live in. They shrug their shoulders in dismay and make comments like, "This house gets pretty cold in the winter" or "This building is pretty drafty." People seem to accept these as immutable facts of life, like "Our bodies deteriorate as we get older." Many people are surprised to learn that there's a lot they can do to seal up the leaks. They're even more surprised to find out that sealing up the leaks in a building envelope can save them a fortune on their energy bills.

How much one can save depends on many factors — most importantly, how leaky the building is. In some buildings, sealing the leaks can reduce heating and cooling costs by 30% or more. They're that leaky! In others, the savings are more modest — around 10%. Even though 10% may not seem like a lot, it can be enormously beneficial over the long haul. If you are spending $1,000 per year to heat and cool your home, a 10% savings is $100 per year. In a decade, a 10% savings amounts to a full year of free heating and cooling. A 30% savings comes to a whopping $300 per year.

Sealing the leaks in a building envelope could cost as little as $50 to $100, so the return on an investment from these measures can be quite substantial. For example, if you spend $100 sealing up the leaks, which saves you $300 per year, your return on investment would be an amazing 300%. If only our retirement or savings accounts performed as well! And, lest we forget, you're not just saving money, you are making your home more comfortable — immediately more comfortable. As a result of these simple changes, you'll be a lot cooler in the summer and a lot warmer in the winter.

## Locating Leaks

The first step to air sealing is to systematically locate the leaks. Once you've found them, you can begin sealing them. All this should take no more than a few hours in most homes, and at most a few hundred dollars in caulk, foam, and weather stripping.

To locate leaks in need of sealing, you can either hire a professional or perform an energy audit yourself.

### *Hiring a Professional*

If you live near a major city, chances are there are at least two or three energy specialists who work in your area. Also called "energy auditors," they perform an analysis of your home energy use — typically referred to as an *energy audit*. You can find a specialist in the Yellow Pages or online under "energy conservation" or "energy auditors." Or, you may want to call a local solar installer or HVAC (heating, ventilation, and air conditioning) company for references.

Energy audits typically cost $300 to $700, depending on the size of your home and how far the auditor has to travel. Energy auditors begin by studying a customer's utility bills. An auditor will probably want to look at your energy bills for at least one to two years. If you don't save your bills, call your local utility and they'll gladly provide the information you need. Or, you may be able to obtain this information on your utility company's website. Using your customer number, you can usually access and download your bills — or a summary of energy consumption. Be sure to have this information ready when the auditor comes, or send it to the auditor's office in advance.

Auditors will note which fuels you use — for example, home heating oil, electricity, propane, or natural gas. They will then examine all loads — that is, all devices like refrigerators, light bulbs, computers, and televisions that consume energy. Auditors often begin by studying the big ticket appliances. They'll note which ones are old and inefficient and therefore in need of replacement. They will take note of unnecessary appliances — like that ancient fridge you hauled to the garage when you bought your new refrigerator. (Second fridges are often energy hogs, especially wasteful when used to house a six-pack or two of soda and a few jars of pickles. And they can cost $200 a year in electricity.)

The auditor will inspect the building envelope inside and out, looking for leaks. He or she will examine the walls, ceilings, floors, foundation, doors, windows, and skylights. During the visual inspection, the auditor searches for obvious leaks in the building envelope. Less obvious leaks will be discovered later by using a special test device.

Leakage occurs through the many cracks and holes in the building envelope, as noted earlier. It also occurs around doors and windows. The older and lower quality the door or window, the more likely it will leak.

Major sources of leakage — and energy waste — are whole-house fans and ceiling fixtures. They typically leak like sieves.

Auditors will also inspect the insulation in a home, approximating the *R-value* of the various parts of the building envelope, like the walls and ceiling. R-value is a measure of the resistance of a material to heat flow; it is primarily determined by the amount of insulation in the building envelope. The higher the R-value, the higher the resistance to heat flow.

Once an auditor has performed a complete visual inspection, he or she typically performs a *blower door test* to quantify leakage and locate the less obvious leaks. Shown in Figure 3-2, the blower door test device is simple, but ingenious, technology that's installed in an exterior door opening. It contains a powerful electric fan that is used to draw air out of the house. (It simulates a 20-mile-per-hour wind blowing on the house.) As air is sucked out of the interior of a building, outside air streams in through cracks and other openings in the building envelope. A digital meter, shown in Figure 3-2, displays the airflow through the building envelope in cubic feet per minute. (A cubic foot of air is about the volume of a basketball.) Once the

PERRY K RK

Fig. 3-2: *This device is used to estimate the infiltration and exfiltration of air and to identify actual leaks.*

test is complete, the auditor converts the air leakage rate under pressure (by the blower door test) to natural air changes — how much room air is replaced naturally by leakage. A rate of around 0.35 air changes per hour or lower is desirable.

To locate the less obvious leaks (of which there are often many), home energy auditors reverse the airflow through the blower door test device. With air blowing *into* a home, the auditor scours the home for leaks using an artificial smoke gun. This hand-held device produces a small stream of artificial smoke. When the device is held close to a leak, the smoke quickly escapes. The auditor will note the presence of these leaks. It's a good idea to follow the auditor around during this process, especially if you intend to seal the leaks yourself.

Air sealing saves in two ways during the winter. In homes heated with forced-air furnaces, air sealing prevents hot air from being forced outside as the furnace blows heated air through the home. The movement of air out of a building is known as *exfiltration*. In most locations in North America, the greatest heat losses occur as a result of this phenomenon.

Sealing leaks also prevents cold air from leaking in on windy days. Air movement into a building is known as *infiltration*. When cold air blows in on the windward side of a house, it enters, then forces warm air out on the downwind side, robbing your home of its heat. This form of heat loss predominates in extremely windy areas, such as eastern Colorado, western Kansas, Nebraska, North Dakota, or the plains of Canada's interior provinces, where winds blow consistently and fiercely throughout the winter.

Sealing the leaks in a building envelope is primarily important in windy areas or for buildings equipped with forced-air furnaces and air conditioning systems (which includes the majority of North America's building stock). Sealing leaks also helps reduce heat loss in buildings that have radiant floor or hydronic heat. Don't skip air sealing just because your home has a radiant floor heating system. And be especially vigilant about sealing leaks high in the building envelope — for example, ceilings in two-story buildings. Warm air builds up in these areas and will leak out through any opening it can find.

After an energy auditor studies your utility bills, examines appliances and the like, determines how leaky the building is, pinpoints all the leaks — both large and small — and determines the R-value of walls, ceilings, and floors, he or she prepares a report. The report should outline recommendations for weatherization (air sealing and insulation) and all other energy improvements such as replacing old, energy-inefficient appliances. These reports typically include the approximate cost and savings for each recommendation. Auditors also usually prioritize their recommendations — that is, they list recommended actions according to their cost effectiveness.

Most energy auditors leave the energy retrofitting to their clients, who can perform the work themselves or hire a professional energy retrofitter to do the job. If you hire a weatherization specialist, they'll need to review a copy of the energy audit, which they will use to propose a plan of attack and determine a price. Once the homeowner and the energy retrofitter have agreed on the scope of the job and the price, the retrofitter will perform the improvements.

If you do nothing else to improve the energy efficiency of your home, be sure to seal up the leaks in the building envelope. This is the most important measure a homeowner can take. Be sure to seal up the leaks *before* you add insulation.

Why seal before you insulate? Because air sealing prevents moisture from entering wall cavities and ceiling insulation. This, in turn, helps to keep insulation dry. Why is that so important? Most forms of insulation experience a dramatic decline in R-value (insulation properties) when wet. In fact, even a tiny amount of moisture can decrease the R-value of insulation by half! So, be sure to seal the leaks in the building envelope first, then install insulation.

Weatherization specialists apply high-quality caulk, spray-in expanding foam insulation, and weather stripping to seal up leaks. It's surprisingly simply and remarkably inexpensive.

### *Locating Leaks Yourself*

An energy audit is one of the most important steps you can take to save energy and reduce your carbon footprint. If the price of a professional auditor is too steep for your pocketbook, however, don't despair. You can perform your own audit to locate leaks and other avenues by which energy is being wasted. To perform a *complete* energy audit, however, you need to have a pretty good understanding of home energy and energy-efficiency measures. If you're experienced in this area, by all means go for it. If not, you can still perform a pretty decent energy audit.

Begin by assessing air leakage. To do so, you'll need to wait for a windy day. (Windy days are more prevalent in the late fall, winter, and early spring in most locations.) Start in the upper floors, then work your way to the main level, then the basement. Begin by looking for the largest and most obvious leaks. Remember: ceiling light fixtures, especially recessed lights, and whole-house fans tend to be major sources of leakage. You can use your hand to feel air leaks, or use a stick of smoldering incense to detect leaks. Air coming in will deflect the smoke stream.

Be thorough. Check the insides of cabinets in bathrooms and kitchens. Check attic accesses and ceiling lights. Check around windows and doors. Be sure to check the junctions where walls meet the ceiling and walls meet floors. Check electrical light switches and electrical outlets on all walls — yes, both inside and outside walls. Large cracks in the basement often occur between the foundation and wall. You can sometimes see light streaming in if you turn the lights off inside.

## Sealing Leaks

As you locate leaks, seal them up. (You may want to enlist the help of another person to follow behind you, sealing leaks you identify.)

Really large cracks — like those between the foundation and sill plate — can be sealed with a foam product generically

DAN CHIRAS

Fig. 3-3: *Backer rod is used to seal large gaps. It seals and insulates and is extremely easy to install.*

known as *backer rod*. It is available in hardware stores and home improvement centers. Backer rod is a solid foam rod that comes in six-foot-long coils. It is pushed into large gaps to block airflow, thereby sealing and insulating.

Many other leaks are sealed with caulk. Caulk comes in three basic types: silicone, silicone/modified polymer, and latex (Figure 3-4). Silicone is the most durable and longest-lasting product and, not surprisingly, the most expensive. The difference in price between silicone and other options is significant, but caulk is still pretty cheap. Shell out a few more bucks per tube to get the best product. You won't regret the decision.

Silicone comes in several colors, but clear and white are the most common. Clear silicone is often used inside the home. It is pretty much invisible (when applied correctly) and will last forever. It expands and contracts without cracking.

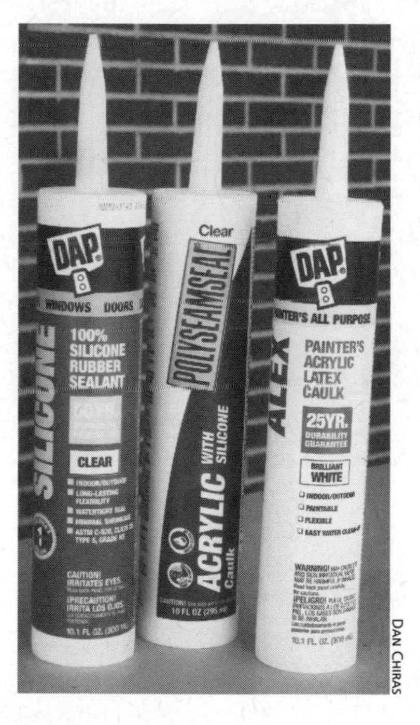


DAN CHIRAS

Fig. 3-4: *Caulk comes in three types: silicone, silicone/modified polymer, and latex. Silicone is generally recognized as the longest lasting.*

To apply silicone, you'll need a caulk gun. Cut the tip of the tube of caulk at a 45-degree angle using a very sharp knife, or better yet, a box cutter or utility knife. Poke a hole through the seal in the tube using the "poker" on the caulk gun. Cut a small portion of the tip off first to see how that works. If the bead is too small, cut a little more off and repeat this until you get it right.

To apply silicone, follow the directions on the tube. Be sure the surface is clean and free of dust. If there's some silicone left in the tube when you're done, don't throw it out. You can insert a three-inch nail through the opening, then seal it with a plastic bag. I use a rubber band to hold the bag in place. You can also seal the tip with electrical tape. Next time you need the caulk, pull out the nail, and you'll be ready to roll.

Cracks can also be sealed with expanding foam sprays (Figure 3-5). Like backer rod, expanding foam seals and insulates (silicone does not). I like working with expanding foam, but it does require a bit of practice and care. Many spray foam products expand at an astonishing rate, creating a cancerous eyesore at the site of application if applied in excess. So, be

DAN CHIRAS

Fig. 3-5: *Expanding foam in spray cans can be used to seal large and small cracks. Small cracks can be filled with Great Stuff's Gaps & Cracks product. Larger cracks can be filled with the appropriately named Big Gap Filler. Small spray cans like the one shown here can be purchased for small jobs, reducing waste.*

sure when starting out to apply expanding foam sparingly. If the foam expands out of the crack, cut the surplus off with a serrated knife or small, sharp saw after it cures. You can then sand and paint the dried foam so it blends in.

Sealing around windows and doors that open and close requires weather stripping, which is designed to seal gaps that can't be sealed shut with caulk or foam. For example, weather stripping is used to seal door jambs (the frames into which doors fit) and operable windows, including double- and single-hung windows and casement windows. Weather stripping comes in many shapes and sizes. Which you use will depend on the job you want it to do.

Most new doors and windows come with weather stripping. But as they age, the weather stripping often cracks or breaks or peels off, requiring replacement.

Self-adhesive foam weather stripping was once commonly sold in hardware and home improvement stores, but, because it doesn't last very long, it has been replaced by more durable products. The same goes for felt weather stripping. Felt weather stripping is tacked in place, but doesn't last as long as newer products and doesn't stop air leakage as effectively.

Most stores sell tubular EPDM rubber weather stripping with a self-adhesive backing. (EPDM stands for *ethylene propylene diene monomer.*) I've been impressed with these products. They go on easily and create a very airtight seal around doors. However, because they're fairly new, I'm not sure how long they'll last. Heavy-duty Duck EPDM rubber Weatherstrip Seal from Henkel Corporation in Avon, Ohio comes with a 10-year guarantee, for what it's worth. M-D Building Products manufactures a tubular vinyl weather strip (Vinyl Gasket) that can be nailed or tacked in place.

Another easy and inexpensive way to seal up the leaks in a building envelope is to install door sweeps on doors and foam gaskets in light switches and receptacle outlets. Door sweeps

DAN CHIRAS

Fig. 3-6: *These thin foam gaskets are installed under the cover plates of electrical outlets and light switches where they block airflow from attics. Air flows into an attic and through walls, then out through light switches and outlets.*

attach to the bottom of doors and block airflow. Foam gaskets for light switches and receptacles are thin pieces of foam that are installed under the cover plates of light switches and electrical outlets. They block air flowing through walls from attics and can be extremely effective (Figure 3-6). Follow the instructions on the package.

For more on caulking and weather stripping, you may want to take a look at my book, *Green Home Improvement*. It includes numerous energy-saving projects and explains most of your options and the associated costs and savings. It also gives lots of advice for installation. Additionally, this book contains detailed instructions on performing home energy audits.

## New Construction

This chapter has focused primarily on existing homes. But air sealing is just as important in new construction, and it's easier to achieve. A good architect can design a building to be pretty airtight. If your builder carefully follows the architect's instructions, your home can be extremely airtight and energy efficient. Concentrating on this at the beginning of a project will save you a fortune over time — and you won't have to seal the building in the future.

To create an airtight building, sill seals should be installed between the foundation and sill plate. The sill plate is the 2 x 4 or 2 x 6 that sits on the top of the foundation. It is part of the wall framing. Sill seals are four-to-six inch rolls of foam applied under 2 x 4s or 2 x 6s, respectively, to block airflow.

Foam gaskets can be applied when installing drywall on walls, too. Be sure to look into airtight drywall installation techniques.

Openings in the building envelope should be carefully sealed. Airtight windows should be installed, as well as airtight insulation-contact recessed light fixtures for recessed lighting.

When building new, be sure to hire an architect *and* a builder who are experienced in designing and building super-airtight, energy-efficient buildings — not just folks who would like to. In other words, find an experienced professional, not just a hopeful. Visit their "end products," too, and be sure to talk with owners and operators of the buildings to see how they've performed.

## Parting Advice

Sealing with caulk and weather stripping is a pretty straightforward process. If hiring a professional, be sure he or she measures air leakage both before *and* after sealing to see how effectively they've sealed up the leaks. Before-and-after tests are vital to assessing the quality of their work and determining if additional measures are required.

If you perform your own weatherization, be sure to continue to check for leaks once you've completed the job. If your home still seems unreasonably cold and drafty, you may want to bring a professional in to find and seal the leaks you've missed. Or, they can locate the leaks using their blower door test device, and you can seal them.

## Conclusion

It may have seemed odd in a book on solar home heating solutions to begin by calling on you to seal up your home before you even begin to think about installing a solar heating system. If you called a HVAC specialist to install a new furnace, he or she would visit your home and jot down a few statistics about your home (the square footage, for example), then size a system. Chances are, he or she would not even mention air sealing or insulation. Unfortunately, that's the wrong approach.

Even if you don't install a solar heating system or retrofit your home for passive solar, it should be clear that air sealing is vital to creating comfortable and affordable living and working spaces. Sealing up the leaks not only makes our buildings much more comfortable, it saves a ton of money while reducing our carbon footprint. If all buildings in the United States, Canada, and Europe — indeed the entire world — were made more airtight, the global community could put a huge dent in their annual emissions of carbon dioxide.

The next step, covered in Chapter 4, is to beef up the insulation. Combined, air sealing and insulation can dramatically lower heat load — the amount of heat you have to provide to stay warm in the winter. And both measures provide a second benefit: they help keep a home cooler in the summer by lowering the cooling load — the amount of energy you spend cooling your home.

Chapter 4

# Insulate, Insulate, Insulate!

When most people think about energy retrofitting their homes or businesses, insulation is the first thought that comes to mind. As you learned in the last chapter, however, for best results, insulation should be the *second* thing on one's mind. The very first measure should be — without an exception — sealing up the leaks.

As you may recall from the last chapter, sealing up the leaks creates a more airtight building that prevents cold air from blowing in on windy days and warm air from leaking out at other times. It makes our buildings much more comfortable year round, and saves a ton of money while reducing one's carbon footprint. Air sealing also helps reduce the accumulation of moisture in insulation, which dramatically lowers its insulating properties. When energy retrofitting a home, insulation should *always* be preceded by air sealing.

Period.

Once you've sealed up the leaks in your home, it is time to insulate.

In this chapter, you will learn how to assess the level of insulation in a building, how much insulation you should add, and which places typically require an insulation shot in the arm. I also discuss common insulation materials and which ones work best in each part of the building envelope. In addition, I cover the costs and benefits of insulation. Although this chapter is primarily concerned with energy retrofitting (boosting the level of insulation in existing buildings), most of these ideas can — and should — be applied to new construction.

## Should You Insulate?

When it comes to insulating a building, the question usually isn't whether you need insulation, but *how much* you need and *where* you need to install it to do the most good. Fact is, most homes and businesses in North America are woefully underinsulated. Millions of buildings throughout North America — indeed the world — would benefit immensely from additional insulation, often lots more!

What is more, an aggressive nationwide weatherization and insulation program could save North Americans billions of dollars a year. It would create an industry with tens of thousands of jobs throughout the Continent that would last for many decades — the need is that great.

Even those who have added insulation in the past decade or so would very likely find that their homes and businesses would benefit from additional insulation, saving them even more on fuel bills. Why?

Until recently the US Department of Energy (DOE) recommended R-30 and R-38 insulation in roofs and ceilings in most parts of the United States. The higher levels (R-38) were recommended for the coldest parts of the United States for homes that were heated with the most expensive type of energy — electricity. The guiding principle behind DOE's insulation guidelines was this: the colder the climate and the more expensive the source of energy for home heating, the higher the level of insulation. Unfortunately, the DOE's recommended R-values that many people have achieved when retrofitting buildings are inadequate.

For years, a small, dedicated handful of building professionals interested in creating superefficient homes has been pushing much higher levels of ceiling insulation — R-50 to R-60 in most climate zones, and R-70 to R-80 in the coldest climates. For walls, they've been advocating R-30 to R-40 in most climate zones, and R-50 or so in the hottest and coldest climates.

Conventionally trained architects and builders often view these more stringent recommendations with skepticism. Like everyone else, they learned in school that there's a point at which additional insulation results in diminished returns. So, they question why one would spend more money for marginal returns.

What architects and builders aren't aware of — because they were never told — is that some forms of insulation lose their R-value over time. Loose-fill insulation (like dry-blown cellulose) may settle over time (Figure 4-1). As it settles, its R-value decreases. In addition, as you know from Chapter 3,

DAN CHIRAS

Fig. 4-1: *Cellulose is made from ground-up newspapers, often with small amounts of cardboard. When blown into attics, this insulation can settle over time, and thus experience a decline in R-value.*

even tiny amounts of moisture in insulation can dramatically lower R-values.

Moreover, the idea of diminishing returns was based on cheap energy — very cheap energy. What architects and builders hadn't been told — or figured out on their own — was that the point of diminishing returns shifts upward as energy prices increase.

For many years, North Americans have been blessed with cheap energy — extremely cheap energy — upon which the recommended insulation levels were based. Those days are gone. Making matters worse, energy prices will very likely continue to increase — and increase quite rapidly — in the coming years. Rising energy prices have dramatically changed the economics of insulation. The old rules no longer apply. More is indeed better — a lot better.

Fortunately, many enlightened architects and builders have adjusted their practices to reflect this new reality. Even the US DOE has quietly increased its recommendations for ceiling insulation from R-30/R-38 to R-49 for most locations, as shown in Table 4-1.

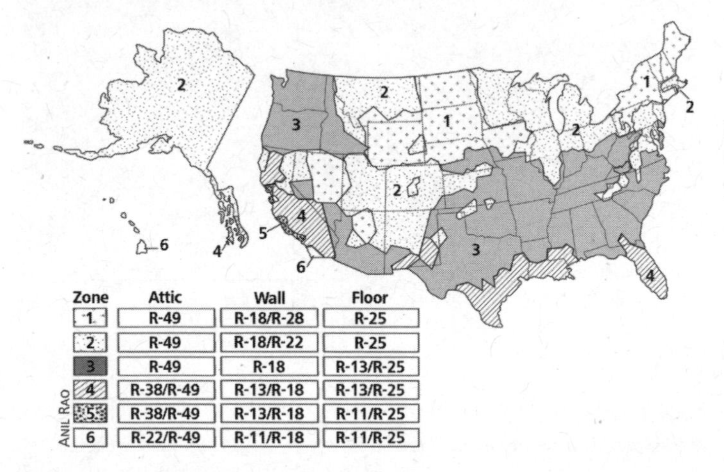


| Zone | Attic | Wall | Floor |
|---|---|---|---|
| 1 | R-49 | R-18/R-28 | R-25 |
| 2 | R-49 | R-18/R-22 | R-25 |
| 3 | R-49 | R-18 | R-13/R-25 |
| 4 | R-38/R-49 | R-13/R-18 | R-13/R-25 |
| 5 | R-38/R-49 | R-13/R-18 | R-11/R-25 |
| 6 | R-22/R-49 | R-11/R-18 | R-11/R-25 |

Table 4-1: *DOE's Recommended R-Values*

While DOE's recommendations for ceiling insulation have increased dramatically, their recommendations for wall insulation are still generally below those that the most energy-conscious green building consultants, architects, and builders recommend (typically R-30 to R-40 in most climate zones). DOE's guidelines also call for R-25 floor insulation over unconditioned spaces, such as crawl spaces. Floor insulation is something builders have usually overlooked.

## How Much Insulation?

When upgrading the insulation in a home or business — or when building a new one — I recommend pushing the limits for ceiling insulation to R-50 to R-60, except in very hot or cold climates, where R-70 to R-80 is advised. Wall insulation for most climates should be R-30 to R-40, except in the hottest and coldest climates, where R-50 is recommended. Floor and slab insulation should be at least R-25.Combined with air sealing, high levels of insulation can reduce heating and cooling costs by as much as 75%.

Contrary to popular belief, high levels of insulation are just as important in hot climates as they are in cold climates. Although conventional wisdom calls for lower levels of insulation in hot climates, in this case conventional wisdom is dead wrong.

Fact is, most of us share the same misconception about insulation. When asked what insulation is for, most people say it is to keep us warm during winter. That's because most people think of insulation as a warm winter coat for their homes. So, it would be useful only during cold weather. You wouldn't bundle up in a heavy coat on a hot summer day, would you?

I remind students that insulation in the building envelope is more like a thermos bottle than a heavy winter coat. If you put a hot beverage in a thermos bottle, it stays hot. If you put a cold drink in the same container, it stays cold.

Insulation reduces heat *movement* through the wall of a thermos bottle the same as it does through roofs, walls, foundations, and windows of buildings. In the winter, it keeps heat in. And in the summer, it keeps heat out. Because cooling requirements (and hence cooling costs) in hot climates often exceed heating requirements (and costs) in cold climates, it is just as important — often more important — to insulate a building in a hot climate as it is in a cold one.

Increasing the R-value of ceiling insulation in a new or existing home to R-50 to R-60 isn't that difficult. For new homes, you can install energy trusses so the insulation levels remain high throughout the ceiling cavity (Figure 4-2). For existing homes, be sure not to block the soffit vents when installing new insulation in the attic.

Adding insulation to an existing building or boosting the amount of insulation of a new building is one of the least expensive upgrades you can make. In fact, in new construction, it could end up being cost neutral initially. That's because increasing the level of insulation often allows one to downsize the heating and cooling system. In an existing home that needs a new furnace, adding insulation may mean you can install a smaller and less expensive unit. The savings created by installing a smaller furnace and air conditioner will easily pay for the additional insulation, making it a cost-neutral action in the short term. Over the long term, higher levels of insulation save money by lowering heating and cooling costs. As a result, they provide a healthy life-long return on the investment.

Insulation upgrades also reduce the cost of solar heating systems, saving money upfront that offsets the cost of the added insulation.

### *Assessing Existing Levels of Insulation*

Before you insulate an existing home, you need to know how much insulation the builder installed initially or how much

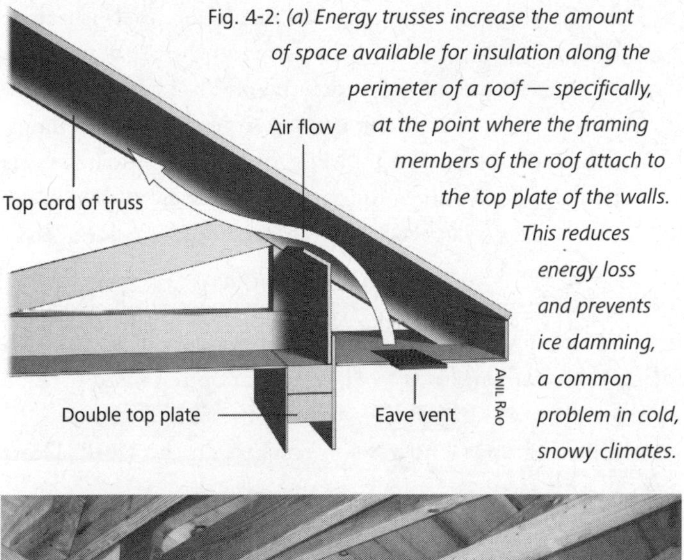


Fig. 4-2: *(a) Energy trusses increase the amount of space available for insulation along the perimeter of a roof — specifically, at the point where the framing members of the roof attach to the top plate of the walls. This reduces energy loss and prevents ice damming, a common problem in cold, snowy climates.*

Fig. 4-2: *(b) Energy trusses in a new building in Wisconsin.*

insulation, if any, previous owners have added. You also need to determine the type and the R-value of the insulation in the walls and ceiling and under floors over unconditioned spaces — that is, spaces that are neither heated nor cooled. (A good example is a floor over a crawl space.) You also need to assess insulation levels around the foundation — either on the outside of the foundation or along the inside of the foundation walls.

One of the easiest ways to assess insulation levels is to hire a professional energy auditor (discussed in the previous chapter). As part of the energy audit, the auditor will determine the type and thickness of the insulation in the walls, ceilings, floors, and foundation of a building. From this data, he or she will calculate the R-values for each part of the building envelope. The auditor will then make recommendations for beefing up the insulation.

Another inexpensive and simple way to determine R-values is to call in a professional insulator. Installers will perform an identical assessment for free, discuss your options, and provide a bid for the work. But please, only call for a bid if you are seriously thinking about hiring someone to do the work. Don't just call for a free assessment; it's not ethical.

A third option is to assess insulation levels yourself. You can, for instance, climb into the attic to measure the depth and determine the type of insulation. You may be able to make your assessment from the safety of a step ladder erected in the attic access. (The attic access is an opening in the ceiling, typically located in a back hallway or closet.) You'll need a flashlight and a ruler or tape measure. You may need to pull yourself into the attic and measure the insulation away from the attic access if the insulation tapers toward the opening, which is common. If you need to climb into the attic to assess the levels, be sure to tread carefully. Step only on ceiling joists, not between them. Once you have your measurement, multiplying the depth of the insulation in the attic by the R-value per inch (Table 4-2) gives the R-value of the ceiling insulation.

In homes with vaulted ceilings, there's no easy way to assess insulation levels. Insulation is packed between framing members (the rafters), and there's no attic to peer into to measure it. I refer to these as *closed ceiling cavities*.

If you live in a home with a vaulted ceiling, one option is to call the original builder to see if he or she can remember how

**Table 4-2**
**R-Values of Insulation**

| Material | R-Value Per Inch | Uses |
|---|---|---|
| **Loose-Fill and Batts** | | |
| Fiberglass (low density) | 2.2 | Walls and ceilings |
| Fiberglass (medium density) | 2.6 | Same |
| Fiberglass (high density) | 3.0 | Same |
| Cellulose (dry) | 3.2 | Same |
| Wet-Spray Cellulose | 3.5 | Same |
| Rock Wool | 3.1 | Same |
| Cotton | 3.2 | Same |
| **Rigid Foam and Liquid Foam** | | |
| Expanded Polystyrene (Beadboard) | 3.8 to 4.4 | Foundations, walls ceilings, and roofs |
| Extruded Polystyrene (Pinkboard and blueboard) | 5 | Same |
| Polyisocyanurate | 6.5 to 8 | Same |
| Roxul (Rigid board made from mineral wool) | 4.3 | Foundations |
| Icynene | 3.6 | Walls and ceilings |
| Air Krete | 3.9 | Walls and ceilings |

much and what kind of insulation was installed. (Good luck with that!) If you were given a set of blueprints, the R-value of the ceiling insulation may be listed on them.

To assess wall insulation, many auditors simply remove the cover plate from an electrical outlet on an exterior wall. This will expose the wall insulation. What they'll typically find is that the cavity is either uninsulated (this is common in homes built in the early 1900s and earlier) or it contains fiberglass batt insulation (batt insulation comes in rolls or precut lengths called "blankets"). Be sure to measure the depth of the wall cavity using a wooden ruler, not a metal ruler or tape measure. (It's not a bad idea to turn off the circuit breaker to that outlet, just in case.) Most older homes were built with 2 x 4s, meaning the wall cavity is about 3.5 inches deep. If your walls were built from

2 x 6s, the wall cavity will be 5.5 inches deep. To determine the R-value of the insulation, multiply the depth of the wall cavity by the R-value per inch of the type of insulation you encounter.

Although this technique often gives a good assessment, many a professional insulation installer has been embarrassed when they showed up with their equipment to inject wall insulation, only to find that the walls were insulated after all. Their cursory test suggested there was no insulation in the walls, but what probably happened was that the electricians who installed the wiring had cut the batt insulation away from the electrical outlets for fire safety and ease of access.

To avoid this embarrassing mistake, you can drill a small hole in an exterior wall using a hole saw (Figure 4-3). This will provide positive proof. (To avoid messing up a wall, work in a closet built against an outside wall. Holes can be easily patched

Fig. 4-3: *To assess wall insulation, you can cut a hole in an outside wall using a hole saw. Drill your hole in an inconspicuous location, like a closet built against an outside wall.*

DAN CHIRAS

up with drywall plugs available from professional installers or a little drywall mud and the circular piece of sheet rock you removed with the hole saw.)

Foundation insulation is pretty easy to assess as well. To see if the foundation was insulated, begin outside. Take a shovel and dig a hole at the foundation. Digging down 6 to 12 inches should be enough. Chances are, you won't find any insulation. Insulating foundations was not a common practice until recently. Even today, many builders fail to insulate foundations.

If your home has a crawl space or a basement, be sure to check for insulation along the inside of the foundation wall. If the basement is finished, you can test for insulation by rapping on the walls as if you were knocking on a door. If the walls are uninsulated, they'll produce a hollow sound. Insulated walls produce a more solid sound. To determine with certainty if the wall cavities are hollow or filled with insulation, you can remove a few cover plates on outside walls or cut a few holes in the wall to look for insulation. For more advice on this topic, you may want to check out my book, *Green Home Improvement,* or some of the other books on home energy efficiency included in the Resource Guide.

## Where to Insulate and What to Use

A careful inspection of insulation performed by an energy professional or on your own will tell you where insulation is needed and how much you must install to properly retrofit your home.

### *Ceiling Insulation*

If the energy audit indicates that you need ceiling insulation, and your home has an unfinished attic, you are in luck. The attic can be accessed via the attic access door and additional insulation is usually very easy to install. Two types of insulation are typically installed in an unfinished attic: loose-fill and batts.

Loose-fill insulation is a dry, fluffy material that's blown into attics. Most loose-fill in use today is either cellulose or fiberglass. It's blown over existing insulation to form a fluffy layer that increases the R-value of your ceiling (Table 4-2). Both products come in tightly packed bales wrapped in plastic. The bales are broken open and fed into the hopper of a blower. Rotating blades in the machine break up the bale and feed the crumbled material into the blower. The blower propels the fluffy cellulose through a large diameter flexible plastic hose that runs to the attic. The insulation is blown over the existing insulation until it reaches the desired depth and R-value.

Batt insulation consists of rolls or blankets cut to predetermined lengths (Figure 4-4). In attics, they can be laid

Fig. 4-4: *Blankets of wool insulation shown here are being installed in a wall cavity in a basement to boost the R-value.*

over the top of existing batt insulation. Batts are available in fiberglass, wool, cotton, and rock wool. Fiberglass is the most widely used and therefore widely available. The most environmentally friendly materials are wool (from sheep) and cotton (made from blue jean factory waste). You'll probably need to purchase wool or cotton batts from an environmental building supply outlet online, like Home Eco in St. Louis or SolSource in Denver.

Batt insulation is typically laid over the top of existing batts. Unfaced batts, that is, rolls or blankets with no paper or foil backing, are typically used when retrofitting attic insulation. They allow moisture seeping up from the living spaces to pass through and escape into the attic and then to the outside via the gable end vents or roof vents. (Venting is vital to ensure that moisture escapes and does not accumulate in attic insulation.)

### *Wall Insulation*

If your energy audit indicates that there's no wall insulation in your home — or wall insulation is inadequate — you have several options. One of the least expensive and most common materials used for insulating uninsulated walls is cellulose. It is blown directly into wall cavities (the spaces between framing members known as "studs") through large holes drilled into the walls, either from the inside or outside of the home. Although you can do this yourself, it's best to hire a professional.

Cellulose is a great product, but it can get hung up in walls during retrofitting. It can, for instance, catch on electrical wires or plaster that protrudes between the wooden lath used in older homes. When cellulose "hangs up," it produces cold spots — uninsulated areas — in the wall.

Walls can also be retrofitted with a product generically referred to as "liquid foam" or "spray foam" insulation. Liquid foam requires an expensive set of equipment that is typically installed in a truck or trailer. To apply this product, you need training on proper mixing and application. As a result, virtually all liquid foam products must be applied by a licensed professional (Figure 4-5). To apply, a hole is typically drilled in the center of the wall. A tube is then snaked into the wall cavity (Figure 4-6), and the liquid foam is injected into the wall. It comes out of the nozzle as a liquid but quickly changes to a foam, which expands to fill the cavity.

Liquid foam insulates and also reduces air infiltration and exfiltration. That is, it not only increases the R-value, it seals up the wall cavities, which dramatically reduces air movement and hence energy loss. To protect your health and the health of other occupants, it's a very good idea to use a product that has no formaldehyde, such as Icynene, Air krete, or BioBased foam (made partly from soybean oil). (Formaldehyde is a suspected carcinogen and can cause other health effects like a

DAN CHIRAS

Fig. 4-5: *Sustainable Energy Systems of Wichita, Kansas, used this spray rig to install liquid foam insulation in the classroom building of The Evergreen Institute. Liquid foam insulation effectively insulates and seals walls, but it must be installed by a professional who has the equipment and training required to do the job right.*

DAN CHIRAS

Fig. 4-6: *The installers inserted a hose in a hole drilled in the wall of our classroom building during its energy retrofit. The hose was slid to the bottom of the wall cavity, and the foam was injected. The hose was drawn out as the foam filled the wall cavity. Once the lower cavity was filled, the hose was snaked to the top, and the rest of the wall was filled from the top down.*

rare, but serious, immune disorder known as *multiple chemical sensitivity*.)

## *Floor Insulation*

If your home is built over a crawl space or an uninsulated basement, be sure to insulate under the floors. Seal up all openings in the floor first, then apply insulation between the framing members (floor joists) upon which the subflooring rests. Floor insulation can make a huge difference in the comfort of a home.

When insulating a floor, you have two options: liquid foam or batts. Liquid foam is sprayed from underneath, filling the space between the floor joists (Figure 4-7). Batts are applied between floor joists and held in place by small metal rods. Be sure to install the paper facing of the batt against the floor to prevent moisture from seeping into the insulation from the

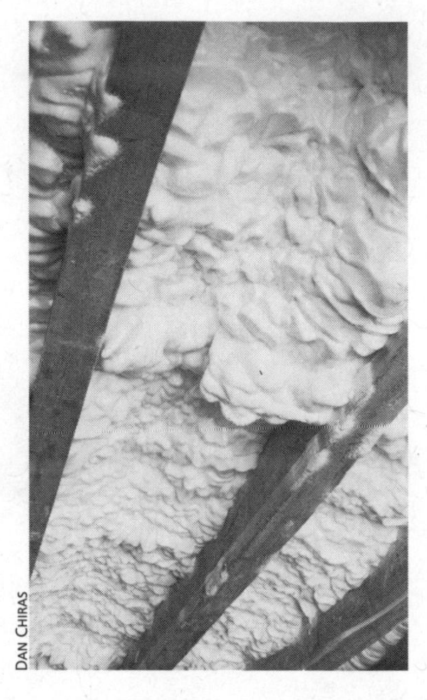


Fig. 4-7: *Liquid foam insulation was installed under the floor of The Evergreen Institute's classroom building during an extensive energy retrofit. The install was donated by one of our collaborators, Sustainable Energy Systems in Wichita, Kansas.*

living area above. When installing insulation in a crawl space, it is often necessary to cover the bare ground with 6 mil polyethylene plastic to reduce the amount of moisture entering the crawl space. This protects the insulation from moisture and helps it retain its R-value.

## *Foundation Insulation*

Uninsulated foundations are a major source of heat loss in homes and businesses. Retrofitting, however, can be quite a challenge. One option is to install rigid foam insulation on the outside of the foundation. To do so, begin by digging a two to four-foot deep ditch alongside the foundation wall. The ditch can be dug by hand or with a back hoe operated by a skilled professional. Rigid foam insulation can then be installed tightly against the foundation wall (Figure 4-8). (If the wall needs additional waterproofing, this is a good time to

DAN CHIRAS

Fig. 4-8: *Rigid foam is used to insulate the exterior walls of foundations. Be sure the product you use is rated for burial or underground applications.*

do it!) The ditch can then be backfilled and the soil will hold the insulation in place.

Be sure to cover exposed (above-ground) insulation, as it will deteriorate in sunlight. Also be sure the insulation you use is rated for burial. Most are, but check just to be sure. Extruded polystyrene is an excellent product. It's also known as *blue board* or *pink board*, and is widely available. Two inches of insulation is typically sufficient, though thicker layers will perform better. Personally, I'd install four inches of rigid foam.

For optimum performance, you may want to install insulation both vertically and horizontally, as shown in Figure 4-9. The horizontal insulation, also known as *wing insulation*, helps trap heat around the foundation that migrates up from the ground on cold winter days. The vertical insulation lowers heat movement through the basement wall. This foundation design, used in many new energy-efficient homes, is known as a *frost-protected shallow foundation*. The design dramatically reduces heat loss, and, for new homes, it reduces the depth of foundations, saving time, energy, and money. (It reduces the depth of the foundation because it raises the frost line around

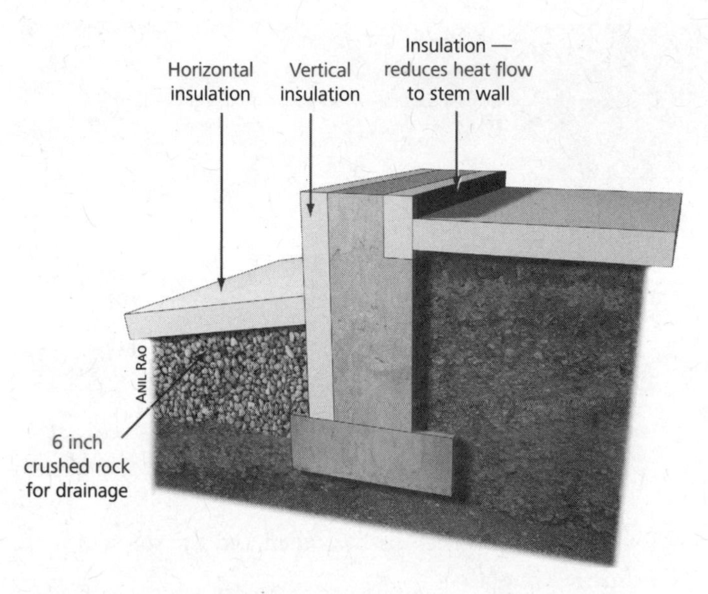


Fig. 4-9: *For best results, apply both horizontal and vertical insulation against your slab or the vertical foundation wall known as a stem wall (if you have a crawl space or basement). This combination reduces heat flow through the foundation wall, greatly reducing heat loss.*

a home. All foundations must be dug deeper than the frost line to prevent expansion and contraction of soils caused by freezing and thawing. Because the frost line is higher around a home with a frost-protected shallow foundation, a shallower and less expensive foundation can be used.)

Horizontal insulation is installed about 18 inches below the surface, over six inches of crushed rock or gravel. The crushed rock or gravel drains moisture away from the foundation wall, helping reduce heat loss. (Wet ground conducts heat away more rapidly than dry soil does.) For advice on insulating your foundation, check out the guidelines published on the National Association of Home Builder's website (www.nahb.org). Search for "frost-protected shallow foundation."

If digging up your foundation doesn't sound like something you want to do, you can also insulate the *inside* surface of foundation walls. This is done by framing the walls with 2 x 4s or 2 x 6s (Figure 4-4). Set the new frame against the concrete wall or even a few inches away to create a thicker wall cavity. Secure the frame to the basement floor using a ramset and the overlying floor joists using nails or screws. Then fill the stud bays (the space between the vertical framing members or studs) with insulation. After the insulation is installed, apply drywall or paneling and finish the walls. Rigid foam insulation can also be applied directly to concrete walls, though you will need a wood framework to attach drywall or paneling, so most people install a batt insulation or fill the cavities with cellulose, fiberglass, or liquid foam.

Never insulate a wall that's wet or damp. Moisture will seep into the insulation, lowering its R-value. It will also very likely cause mold to form in the insulation. Moisture, if severe enough, can cause the framing members to rot.

So, if your basement walls leak, fix that problem first — *before* you insulate. To reduce moisture seeping into basement walls, you may need to install gutters and downspouts, if you don't have any, or replace or repair leaky gutters. Downspouts should carry water at least six feet away from the foundation — although ten feet is even better. Also, be sure to grade the soil around your home so it slopes away from the foundation. Other, more drastic measures are sometimes required. When in doubt, call in a professional builder to provide advice. You can also check out my book, *Green Home Improvement,* which offers tons of advice on reducing water leakage in basements.

## Conclusion

Air sealing and insulation can save you a fortune over the long run, slashing heating costs in half — or even more. They'll also reduce the initial cost of a new building. (You'll need a much smaller heating and cooling system.) At the very least, these measures can be cost neutral. However, because they save on energy bills in the summer and winter, they provide long-term economic savings, all while creating a more comfortable living or working space.

When building anew, you may end up arguing a bit with your architect or builder about your desire to install such high levels of insulation. As always, it is best to hire someone who sees eye to eye with you — someone who already designs and builds superinsulated buildings. Be wary of inexperienced architects or builders who promise you anything to get the job, but quickly disagree with you over high levels of insulation after the contract is signed.

If you are retrofitting a house, you may even find yourself spending a considerable amount of time convincing the insulation installer of the merits of higher-than-conventional levels of insulation. Be persistent. They're often steeped in the old logic. You may have to invoke the Nike mantra, "Just do it!" After all, it is your home — and your dime.

CHAPTER 5

# UNDERSTANDING SOLAR ENERGY

Now that you know the most important things you need to do to prepare your home for a solar heating system, we can turn our attention to the source of energy you'll be tapping: the Sun. In this chapter, I'll explore the Sun and solar energy, providing you with the information you need to understand to get the most out of a solar home heating system.

## Understanding Solar Radiation

The Sun lies in the center of our solar system, approximately 93 million miles from Earth. Composed primarily of hydrogen and small amounts of helium, the Sun is a massive fusion reactor. In the Sun's core, intense pressure and heat force hydrogen atoms to unite, or fuse, creating slightly larger helium atoms. In this process, immense amounts of energy are released; this energy migrates to the surface of the Sun and then radiates out into space, primarily as light and heat.

A small portion of this light and heat streaming through space strikes the Earth, warming and lighting our planet, and fueling aquatic and terrestrial ecosystems. According to French energy expert, Jean-Marc Jancovici, "the solar energy received each year by the Earth is roughly ... 10,000 times the total energy consumed by humanity." To replace *all* the oil, coal, gas, and uranium currently used to power human society with solar energy, we'd need to capture a mere 0.01% of the energy of the sunlight striking Earth each day.

The Sun's output is known as *solar radiation*. As shown in Figure 5-1, solar radiation ranges from high-energy, short-wavelength gamma rays to low-energy, long-wave radiation known as *radio waves*. In between these extremes, starting from the short-wave end of the spectrum are x-rays, ultraviolet radiation, visible light, and heat (infrared radiation).

While the Sun releases numerous forms of energy, most of it (about 40%) is infrared radiation (heat) and visible light (about 55%). Traveling at 186,000 miles per second, solar energy takes 8.3 minutes to make its 93 million mile journey from the Sun to the Earth. Solar radiation travels virtually unimpeded through space until it encounters Earth's atmosphere. In the outer portion of the atmosphere (a region known as the *stratosphere*) ozone molecules ($O_3$) absorb much (99%) of the incoming ultraviolet radiation, dramatically reducing our exposure to this potentially harmful form of solar radiation. As sunlight passes through the lower portion of the atmosphere (the troposphere), it encounters clouds, water vapor, and dust. These may either absorb some of the Sun's rays or reflect them back into space, reducing the amount of sunlight striking Earth's surface. Absorbed solar radiation in the visible range is converted into heat.

Solar heating systems — including passive solar homes, solar hot air systems, and solar hot water systems — capture the energy contained in the visible and lower end of the infrared portions of the spectrum, known as *near-infrared radiation* (Figure 5-1).

### *Irradiance*

The amount of solar radiation striking a square meter of Earth's atmosphere or Earth's surface is known as *irradiance*. It

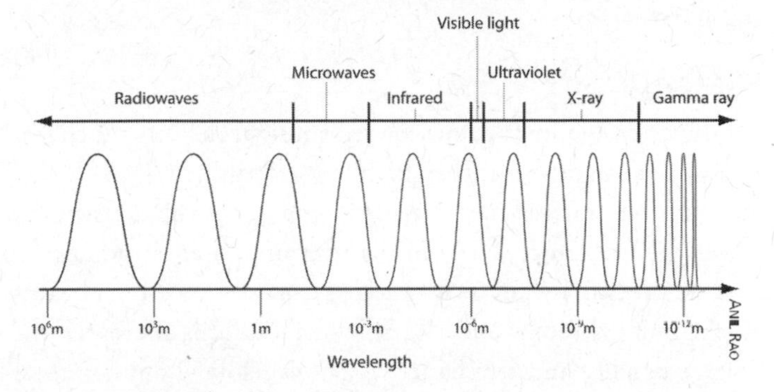


Fig. 5-1: *The top bar shows the wavelengths of the different forms of radiation emitted by the Sun. The lower bar shows the visible and near-infrared portions of the spectrum — the portion that provides us with the energy we use to heat our homes. Much of the ultraviolet radiation is blocked by window glass.*

is measured in watts per square meter ($W/m^2$). Solar irradiance measured just before it enters the Earth's atmosphere is about 1,366 $W/m^2$. On a clear day, nearly 30% of the Sun's radiant energy is either absorbed and converted into heat or reflected by dust and water vapor back into outer space. By the time the incoming solar radiation reaches a solar collector on a roof, the incoming solar radiation is reduced to about 1,000 $W/m^2$.

Solar irradiance varies during daylight hours at any given site. At night, solar irradiance is zero. As the Sun rises, irradiance increases, peaking around noon. From noon until sunset, irradiance slowly decreases, falling once again to zero at night. These changes in irradiance are determined by the angle of the Sun's rays, which changes continuously as the Earth rotates on its axis. The angle at which the Sun's rays strike Earth affects both the energy density (Figure 5-2) and the amount of atmosphere through which sunlight must travel to reach Earth's surface (Figure 5-3). Both affect the daily output of solar heating systems.

## ENERGY DENSITY

As shown in Figure 5-2, low-angled sunlight delivers much less energy per square meter than high-angled sunlight. Low-angled sun thus has a *lower energy density*. The lower the density, the lower the irradiance. Early in the morning, then, irradiance is low. As the Sun makes its way across the sky, however, it beams down more directly onto Earth's surface. This increases the energy density and irradiance. A passive solar home or solar collector installed for space heating will therefore gain more solar energy as the day progresses toward noon (and slightly after noon).

Irradiance is also influenced by the amount of atmosphere through which the sunlight passes, as shown in Figure 5-3. The more atmosphere through which sunlight passes, the more filtering occurs. The more filtering, the less sunlight makes it to Earth. The less sunlight, the lower the irradiance. Solar space-heating systems must be oriented to capture as much sunlight as possible during this period of lowest irradiance — that is, during the heating season, when the Sun is low in the sky.

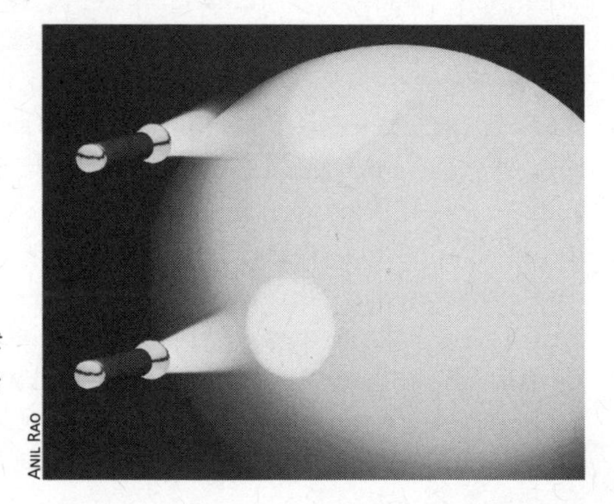


Fig. 5-2: *Surfaces perpendicular to incoming solar radiation absorb more solar energy than surfaces not perpendicular. This has many important implications when it comes to mounting solar collectors.*

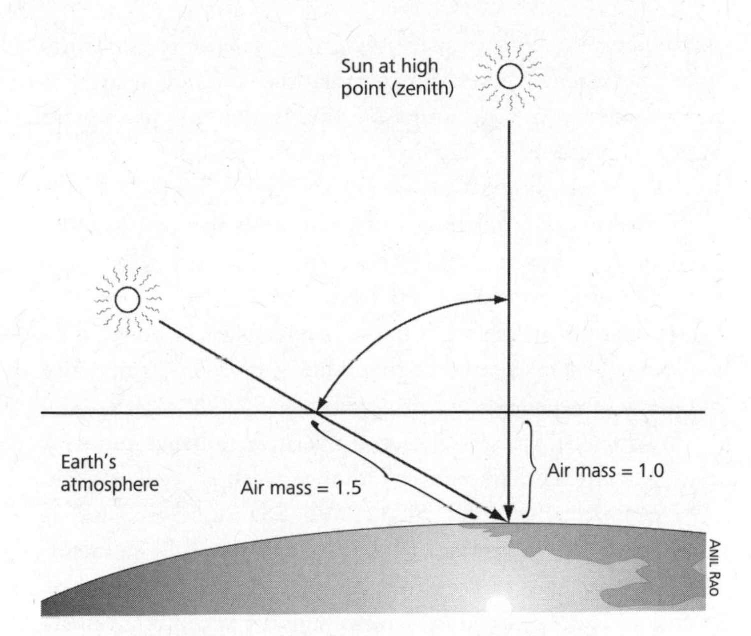


Fig. 5-3: *Early and late in the day, sunlight passes through more atmosphere. This reduces irradiance. Maximum irradiance occurs when the Sun's rays pass through the least amount of atmosphere, that is, at solar noon (halfway between sunrise and sunset). Three fourths of a solar space-heating system's output occurs between 9 am and 3 pm each day.*

## *Irradiation*

Irradiance is an important measurement, but what most solar installers need to know is *irradiance over time* — that is, the amount of energy that will strike a solar collector or enter a passive solar home in a 24-hour period. Irradiance over a period of time is referred to as *solar irradiation.* It's expressed as watts per square meter striking Earth's surface (or a PV module) for some specified period of time — usually an hour or a day. Hourly irradiation is expressed as watt- hours per square meter. (Remember watt-hours and kilowatt-hours discussed in Chapter 2?) For example, solar radiation of 500 watts of solar energy striking a square meter for an hour is 500 watt-hours per square meter. Solar radiation of 1,000 watts per square meter over two hours is 2,000 watt-hours per square meter. To determine watt-hours, simply multiply watts per square meter by hours.

To help keep irradiance and irradiation straight, you can think of irradiance as a measure of instantaneous power (measured in watts). Irradiance is therefore a *rate*.

Irradiation, on the other hand, is a measure of power during some specified period of time and is, therefore, a measure of energy. It is a *quantity*.

Teachers help students keep the terms irradiance and irradiation straight by likening irradiance to the speed of a car. Like irradiance, speed is an instantaneous measurement. It simply tells us how fast a car is moving at a particular moment. Irradiation is akin to the distance a vehicle travels. Distance, of course, is determined by multiplying the speed of a vehicle by the time it travels at a given speed. In a car, the longer you travel, the greater the distance you'll cover. In solar energy, irradiation increases with time.

Figure 5-4 illustrates the concepts graphically. In this diagram, irradiance is the single black line in the graph — the number of watts per square meter that strike a surface at any moment in time. The area under the curve is solar irradiation — the total solar irradiance over time. In this graph, it's the irradiance occurring in a day. Solar irradiation is useful in sizing all types of solar systems.

## The Sun and the Earth: Understanding the Relationships

Now that you understand irradiance and irradiation, let's examine the geometric relationships between the Earth and Sun. An understanding of the ever-changing relationship between the Earth and the Sun helps you understand how

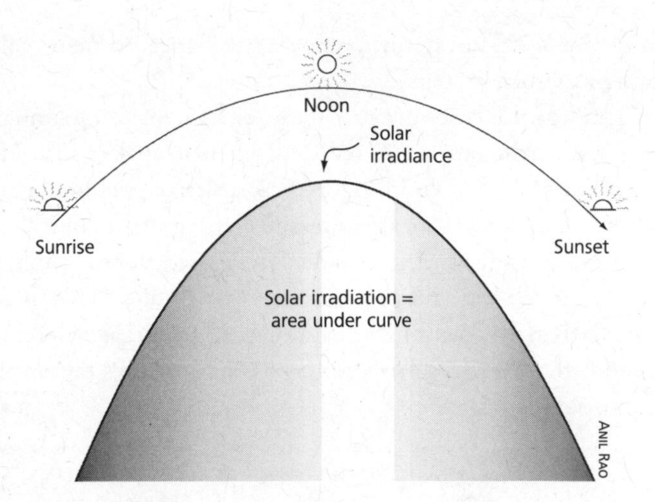


Fig. 5-4: *This graph shows solar irradiation, watts per square meter, and irradiation in watts per square meter in a day. Irradiation is the area under the curve.*

passive solar systems work and the best position (orientation and angle) for a solar hot water or solar hot air collector.

## *Day Length and Altitude Angle: The Earth's Tilt and Orbit Around the Sun*

As you learned in grade school, the Earth orbits around the Sun, completing its path every 365 days (Figure 5-5).

As shown in Figure 5-6, the Earth's axis is tilted 23.5°. The Earth maintains this angle as it orbits around the Sun. Look carefully at Figure 5-5 to see that the angle remains fixed — almost as if Earth were attached to a wire attached to a fixed point in outer space. Because Earth's tilt remains constant, the Northern Hemisphere is tilted away from the Sun during its winter (Figure 5-5). As a result, the Sun's rays enter and pass through Earth's atmosphere at a very low angle. Sunlight penetrating at a low angle passes through more atmosphere, where it is absorbed or scattered by dust and water vapor, as shown in Figure 5-3. This, in turn, reduces irradiance, reducing solar gain by all solar systems.

Irradiance is also lowered because the density of sunlight striking Earth's surface is reduced when it strikes a surface at an angle. As shown in Figure 5-2, a surface perpendicular to the Sun's rays absorbs more solar energy than one that's tilted away from it. As a result, low-angled winter sunlight delivers much less energy per square meter of surface in the winter than it does during summer. During the winter in the Northern Hemisphere, most of the Sun's rays fall on the Southern Hemisphere.

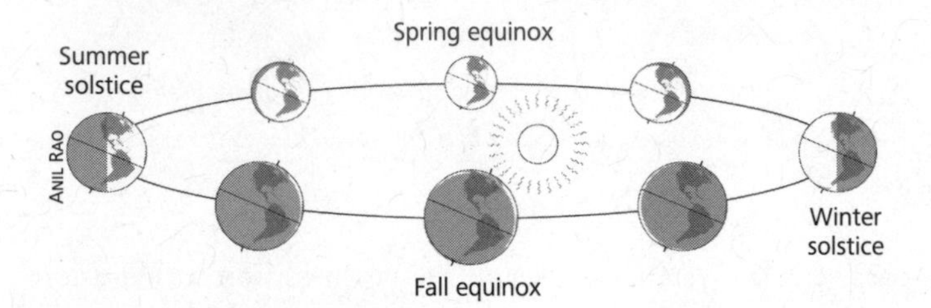


Fig. 5-5: *Note that Earth is closest to the Sun in the winter in the Northern Hemisphere, but the Northern Hemisphere is tilted away from the Sun, so the Sun's rays penetrate the atmosphere at a low angle.*

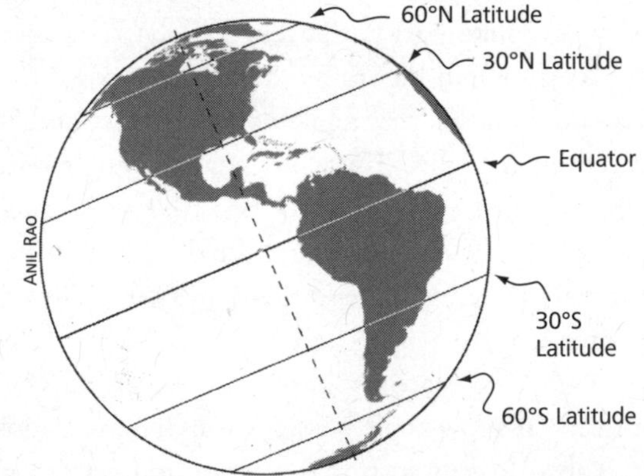


Fig. 5-6: *The Earth is tilted on its axis of rotation, a simple fact with profound implications.*

Solar gain is also reduced in the winter because days are shorter — that is, there are fewer hours of daylight. Day length is determined by the angle of the Earth in relation to the Sun.

In the summer in the Northern Hemisphere, the Earth is tilted *toward* the Sun, as shown in Figures 5-5 and 5-7. This results in several key changes. One of them is that the Sun is positioned higher in the sky. As a result, sunlight streaming onto the Northern Hemisphere passes through less atmosphere, which reduces absorption and scattering. This, in turn, increases solar irradiance. Because a surface perpendicular to the Sun's rays absorbs more solar energy than one that's tilted away from it, the Northern Hemisphere intercepts more

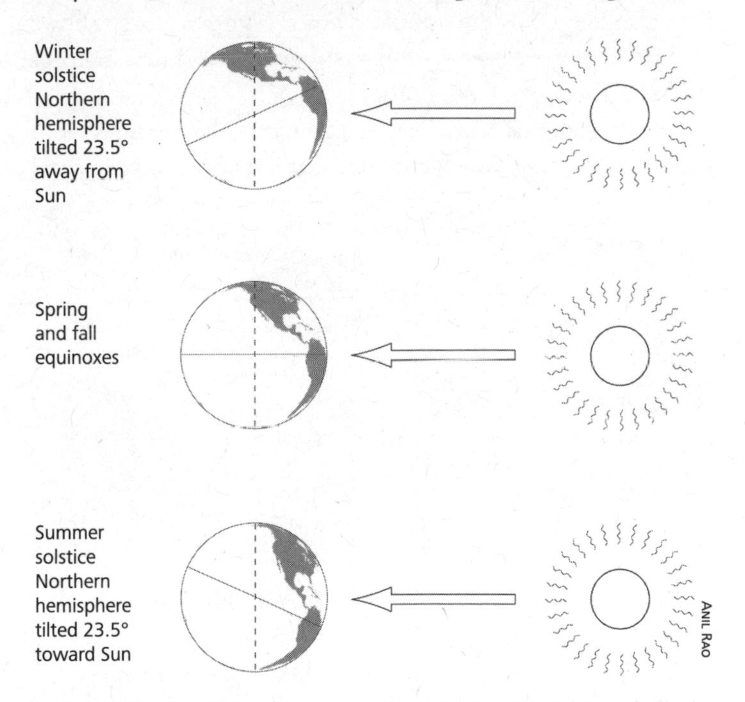


Fig. 5-7: *Note that during the summer, the Northern Hemisphere is bathed in sunlight. The Sun's rays enter at a steep angle. In the winter, the Sun's rays enter at a low angle.*

energy during the summer. Put another way, high-angled Sun delivers much more energy per square meter of surface area than in the winter. Moreover, days are longer in the summer.

Keep these facts in mind as you read about the three solar home heating options I'll be covering: passive solar, solar hot air, and solar hot water. They're crucial concepts one must know to optimize the performance of each system.

Figure 5-8 shows the position of the Sun as it "moves" across the sky during different times of the year as a result of the changing relationship between the Earth and the Sun. As just discussed, the Sun "carves" a high path across the summer sky. It reaches its highest arc on June 21, the longest day of the year, also known as the *summer solstice*. Figure 5-8 also shows that the lowest arc occurs on December 21, the shortest day of the year. This is the *winter solstice*.

For curious readers, the English word *solstice* comes from the Latin *solstitum*, which comes from *sol* for Sun and *ìsunî* and

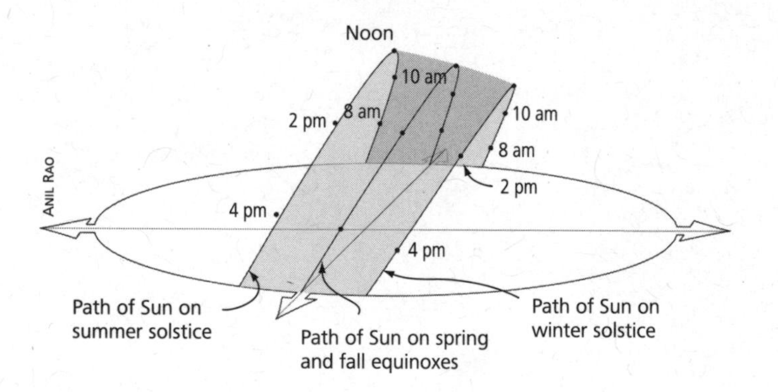


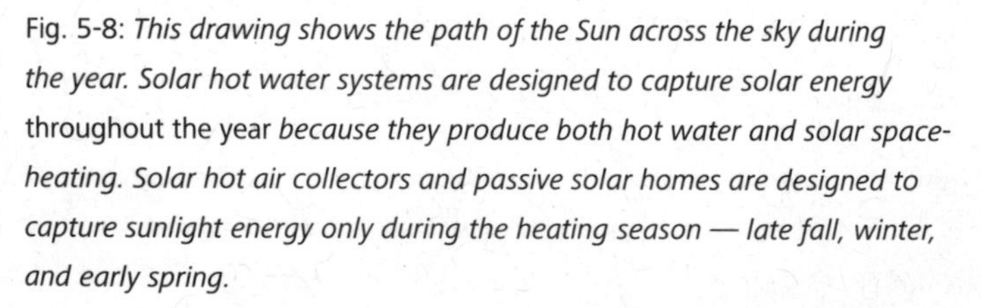

Fig. 5-8: *This drawing shows the path of the Sun across the sky during the year. Solar hot water systems are designed to capture solar energy* throughout the year *because they produce both hot water and solar space-heating. Solar hot air collectors and passive solar homes are designed to capture sunlight energy only during the heating season — late fall, winter, and early spring.*

*-stitium*, meaning a stoppage. The summer and winter solstices are those days when the Sun halts its upward and downward "journey."

The angle between the Sun and the horizon at any time during the day is referred to as the *altitude angle*. As shown in Figure 5-8, the altitude angle decreases from the summer solstice to the winter solstice. After the winter solstice, however, the altitude angle increases, growing a little each day, until the summer solstice returns. Day length changes along with altitude angle, decreasing for six months from the summer solstice to the winter solstice, then increasing until the summer solstice arrives once again.

The midpoints in the six-month cycles between the summer and winter solstices are known as *equinoxes*. The word *equinox* is derived from the Latin words *aequus* (equal) and *nox* (night). On the equinoxes, the hours of daylight are nearly equal to the hours of darkness. The spring equinox occurs around March 20, and the fall equinox occurs around September 22. These dates mark the beginning of spring and fall, respectively.

As just noted, the altitude angle of the Sun is determined seasonally by the angle of the Earth in relation to the Sun. The altitude angle is also determined daily by the rotation of the Earth on its axis. As seen in Figure 5-8, the altitude angle increases between sunrise and noon, then decreases to zero once again at sunset.

The Sun's position in the sky relative to a fixed point, such as a solar hot water system collector, also changes by the minute. Solar installers locate the Sun's position in the sky in relation to a fixed point by using the *azimuth angle*. As illustrated in Figure 5-9, true south is assigned a value of 0°. East is +90° and west is -90°. North is 180°. The angle between the Sun and 0° south (the reference point) is known as the *solar azimuth angle*. If the Sun is east of south, the azimuth angle falls in the range of 0 to +180°; if it is west of south, it

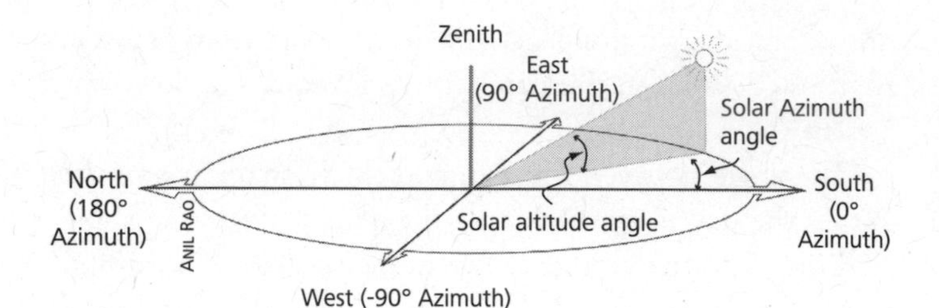


Fig. 5-9: *The altitude angle is the angle between the horizon and the Sun. The azimuth angle is the angle of the Sun in relation to true south, which is assigned a value of zero degrees. Both the azimuth and altitude angles change throughout the day as the Sun "moves" across the sky.*

falls between 0 and -180°. Like altitude angle, azimuth angle changes as a result of the Earth's rotation.

## *Implications of Sun-Earth Relationship for Solar Heating System*

In the chapters on passive solar, solar thermal, and solar hot air systems that follow, you'll learn how the information presented in this chapter affects installations. For now, it is important simply to note that for optimum solar gain, passive solar homes and collectors in solar thermal and solar hot air systems should be oriented to capture as much of the low-angled winter sun as possible for optimum performance. They must also be oriented to true south (in the Northern Hemisphere), not magnetic south. What's the difference?

True north and south are measurements used by surveyors to determine property lines. They are imaginary lines that run parallel to the lines of longitude, which, of course, run from the North Pole to the South Pole. (True north and south are also known as true *geographic* north and south.)

Magnetic north and south, on the other hand, are determined by Earth's magnetic field. They are measured by compasses. Unfortunately, magnetic north and south rarely line up with the lines of longitude — that is, they rarely run true north and south. In some areas, magnetic north and south can deviate quite significantly from true north and south. How far magnetic north and south deviate from true north and south is known as the *magnetic declination.*

Figure 5-10 shows the extent of the deviation of magnetic north and south — the magnetic declination — from true north and south in North America. Take a moment to study the map. Start by locating your state and then read the value of the closest isobar. If you live in the eastern United States, you'll notice that the lines are labeled with a minus sign. This indicates a westerly declination. What this means is that true south is located west of magnetic south. If you live in the midwestern and western United States, the lines are positive. This indicates an easterly declination. Again, that means that true south lies east of magnetic north and south.

To determine the magnetic declination for where you live, you can consult this map, although a local surveyor or nearby

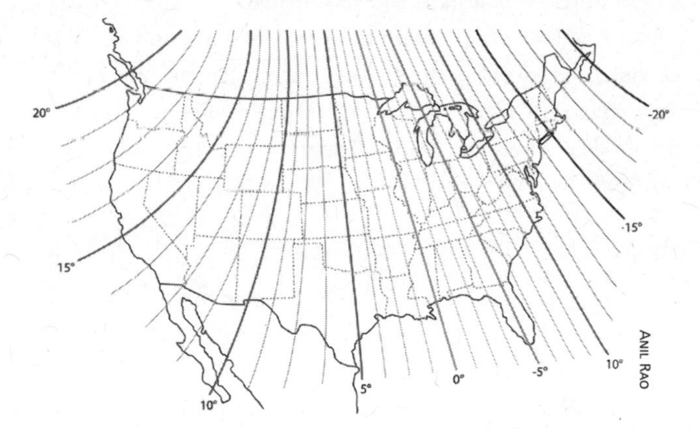


Fig. 5-10: *The isobars on this map indicate magnetic declination in the United States. Note that negative numbers indicate a westerly declination (meaning true south lies west of magnetic south). Positive numbers indicate a easterly declination (meaning true south is east of magnetic south).*

airport can provide a more precise reading. Be sure to ask whether the magnetic declination is east or west. Bear in mind that magnetic north and south not only deviate from true north and south, they also change very slightly from one year to the next. Surveyors keep track of the annual variation. Also be aware that compass readings at any one site may be slightly off. If you take a magnetic reading too close to your vehicle, for instance, the compass may not point precisely to magnetic south.

Orienting a solar system to true south is ideal, but it's not always possible. Don't sweat it. You have some leeway — especially when it comes to solar thermal (hot water) and solar hot air systems. In fact, orienting a solar hot air collector and solar thermal collector system slightly off true south results in only a very slight decrease in output until the azimuth angle is off by about 25°. As you will see in Chapter 9, orienting a solar home slightly off true south has very little effect on solar gain in the winter, but it can result in a substantial — and unwanted — increase in solar heat gain in the summer.

The angle at which solar hot air collectors and solar thermal collectors intended for space heating are mounted is set to ensure maximum solar gain in the winter when heat is needed most — also when the Sun is lowest in the sky. (More on this in subsequent chapters.)

## Conclusion

Solar energy is an enormous resource that could provide heat to millions of homes throughout the world and help us reduce our carbon footprint and eliminate many of the socially, economically, and environmentally costly impacts of our heavy reliance on fossil fuels. Solar heating systems can be very cost effective and almost always represent a wise decision. You must, however, orient these systems correctly to maximize solar gain.

# CHAPTER 6

# PASSIVE SOLAR HEATING: LOW-TECH, HIGH PERFORMANCE

Each day, the low-angled winter sun heats millions of buildings in the Northern Hemisphere. Streaming in through south-facing windows, the Sun's rays warm the interiors of homes and offices. Most of this solar heating is not intentional, however. That is, most of these homes are not designed to capture the Sun's energy. It just happens. Their south-facing windows simply allow solar energy from the low-angled winter sun to pass through them (provided the windows are not covered by curtains or blinds).

Among the legions of accidentally solar-heated homes are many intelligently designed and carefully oriented homes that intentionally capture solar energy to provide wintertime heat (Figure 6-1). These are known as *passive solar homes*. "Passive" refers to the fact that these buildings do not require complicated or costly mechanical systems to generate and distribute heat (though they may be needed for backup heat). A passive solar home is as low-tech as you can get. They're inexpensive to build and easy to operate.

This chapter discusses the design and construction of a passive solar home — actually any building that is heated by the Sun. These ideas generally apply to new construction. If you are retrofitting a house for passive solar, the next chapter

DAN CHIRAS

Fig. 6-1: *This home designed by James Plagmann of HumaNature Architecture and me allows the low-angled winter sun to enter through south-facing windows.*

discusses ways you can do that. Be sure to read this chapter first, however, as it provides important concepts you'll need to know.

Passive solar design can save you thousands of dollars a year — and tens, even hundreds, of thousands of dollars — over a period of 20 to 30 years. The savings to the owner are substantial enough to pay for a child's four-year college education!

In the following sections, I'll introduce the basic principles of passive solar design. I will then discuss how these principles are used to create passive solar buildings.

## Principles of Passive Solar Heating

Passive solar heating relies on a handful of time-tested design principles to ensure its success. Study them carefully, implement them in your building design, and you'll be rewarded with a lifetime of free heat.

### *Orient to True South*

First and foremost, a passive solar home must be oriented to true south (in the Northern Hemisphere). What that means is that the long axis of the home runs from east to west. This results in a rather large south-facing surface in which windows can be installed to permit the low-angled winter sun to enter.

A solar home should be oriented to true south, not magnetic south. As noted in the last chapter, true north and south are the lines of longitude that run from the North Pole to the South Pole. Magnetic south is detected by a compass, but rarely corresponds to true south.

In fact, as you can see in Figure 6-2, magnetic lines deviate substantially from true north and south. This phenomenon, discussed in Chapter 5, is referred to as *magnetic declination.* As illustrated, the only place where magnetic south and true south correspond in North America is near the very center of the Continent. At our educational center in east-central Missouri, for instance, magnetic south is only about one degree off from true south. Travel to St. Louis, about 70 miles east of our campus, and you'll find that magnetic and true south line up perfectly.

As shown in Figure 6-2, magnetic declination west of the "zero line," which runs through the center of the Continent, increases. If you were to design a passive solar home in

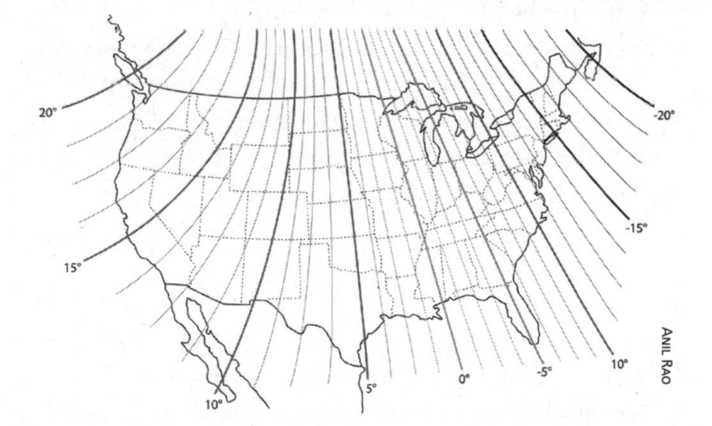


Fig. 6-2: *This map shows the difference between magnetic and true south.*

east-central Kansas, the compass reading would be 5° off. Colorado would be about 10° off. But which way? Is true south east or west of magnetic south?

When designing a passive solar home located in a site west of the "zero line," true south lies east of magnetic south. In a site in Kansas, then, magnetic south would be 5° east of magnetic south. In eastern Wyoming, central Colorado, or north central New Mexico, true south would about 10° east of magnetic south. (Take a look at the line in Figure 6-2 to confirm this assertion.) If you were siting a home in western Montana or eastern Idaho, true south would be 15° east of magnetic south.

East of the "zero line," magnetic declination is westerly. If you are siting a home in central Florida, for instance, true south would be 5° west of magnetic south. In western Pennsylvania, true south would be 10° west of magnetic south.

Don't forget to adjust for magnetic declination! If this detail slips your mind, your passive home will collect a lot less sunlight energy in the winter and too much in the summer, causing overheating. The goal, once again, is to orient a solar home as close to true south as possible to maximize solar gain.

### *Concentrate Windows on the South*

Passive solar homes and businesses are really large, building-sized solar collectors. They are designed to capture the Sun's energy and turn it into heat. Like other types of solar collectors, solar homes require a transparent surface to allow sunlight in. In the case of a solar home, the transparent surfaces are the south-facing windows. Sunlight streams through this glass (called *solar glazing*), warming the interior.

To ensure maximum solar gain, windows in a passive solar home are concentrated along the southern wall, where they'll do the most good on a cold winter day. My rule of thumb is that the square footage of solar glazing should be equal to 12% to 15% of the square footage of the building (for a one-story building with a conventional ceiling height of eight or nine feet). Higher levels may be required to increase solar gain. Remember that window surface area is only the area occupied by glass, not the frames of windows. To determine the square footage of glass, designers typically multiply the window dimensions to determine total square footage, then deduct 25% to take the framing into account.

To minimize heat loss, north-facing windows should be limited to no more than 4% of the total square footage. East-facing windows are limited to 4% of the square footage of the building, and west-facing windows are limited to 2% to reduce heat gain in the summer. As an example, a 1,000-square-foot single-story building would require 120–150 square feet of south-facing glass, but no more than 40 square feet of north- and east-facing glass, and no more than 20 square feet of west-facing glass.

These rules are general design guidelines for the ideal passive solar home. Most designs violate these rules for one reason or another. If a home faces a beautiful lake on the west side, for example, an owner might want to install a few more windows on that side to take in the view. To prevent overheating, however, the owner would need to install special glass to reduce heat gain in the summer through the west-facing windows. This glass needed for these windows is known as *low solar heat gain coefficient* (SHGC) *glass*.

### *Create an Airtight, Energy-Efficient Building*

Another key to successful passive solar design is to design and construct buildings that are airtight and superinsulated. As you learned in Chapters 3 and 4, these measures help retain heat inside a home. Because solar energy is a rather diffuse (low-concentration) form of energy, you have to use every trick in the book to gather up and retain solar heat. There's no sense letting that hard-won solar energy escape through a leaky, poorly insulated building envelope.

For more on airtight energy-efficient design, be sure to read Chapters 3 and 4, if you haven't already done so. If you want to learn more about insulation and air sealing, look at my book, *Green Home Improvement* or *The Solar House: Passive Heating and Cooling*. The latter provides detailed coverage of this topic. (For those who speak Chinese, there's now a Chinese edition!)

## *Incorporate Thermal Mass*

Passive solar buildings absorb sunlight during daylight hours. This heat warms the interiors, providing comfort. To heat a building at night, designers incorporate *thermal mass* in their buildings (Figure 6-3). Thermal mass consists of any solid masonry-type material that absorbs heat. It acts like a heat sponge, absorbing warmth during the day, then releasing it at night or on cold, cloudy days. Concrete and tile floors in solar homes and brick or stone-faced interior partition walls strategically located in a passive solar home can often be employed as thermal mass — if properly placed. They'll release heat at night and on cloudy days if they've been charged by the Sun.

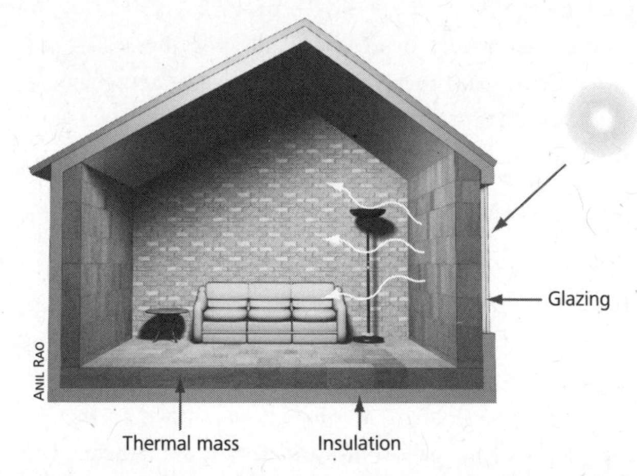


Fig. 6-3: *This diagram shows thermal mass in a passive solar home that absorbs heat during the day, releasing it at night or on cloudy days.*

Thermal mass is generally dark-colored mass, a feature that increases its absorption capacity — but not so dark as to create hot spots. Thermal mass should be about four inches thick for optimum performance and is best located so it is directly in the path of the incoming solar radiation (Figure 6-4). This usually

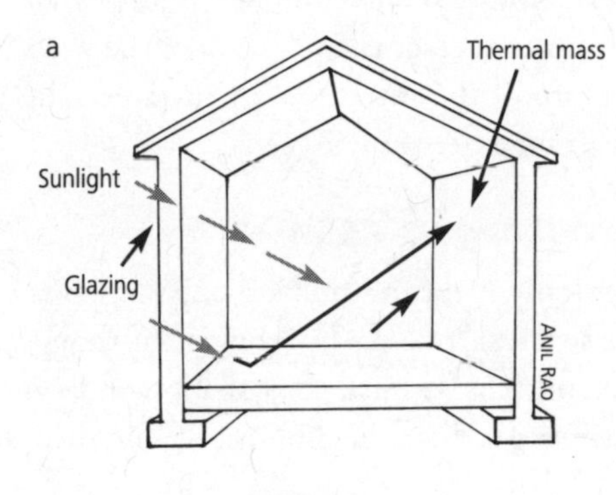


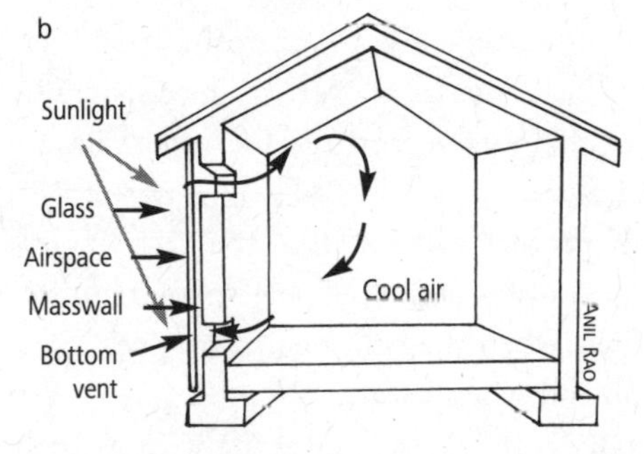


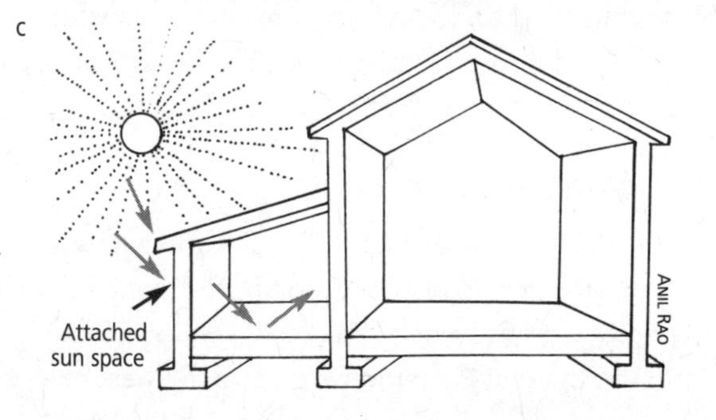


Fig. 6-4: a-c *This drawing shows the optimum placement of thermal mass in the direct path of incoming solar radiation. Thermal mass located in the back walls is also vital to achieving maximum comfort.*

means that most of the thermal mass should be located on the south side of the home's interior, near the solar glazing.

For maximum comfort, however, thermal mass should surround the living space. That way, heat absorbed during the day will radiate from all interior wall surfaces at night, making rooms feel warmer. (Comfort is determined not just by air temperature but also by mean surface temperature of the walls — that is, the temperature of the interior wall surfaces. The warmer they are, the warmer a person will feel.)

### *Provide Adequate Overhang*

Passive solar homes typically incorporate overhangs or eaves, at the very least, on the south side of the building. (Overhangs are a good idea in any building because they shade windows and walls in the summer and reduce the amount of rain dripping down walls.)

Overhangs shade the solar glazing in late spring, summer, and early fall, preventing buildings from overheating. Generally, passive solar homes require a two- to three-foot overhang to protect against the summer sun.

Overhangs not only protect a home from the high-angled summer sun, they are the on-off switches of passive solar homes. The overhang determines when the Sun begins to "peek" into the windows in the fall, and when it can no longer gain access in the spring. As you may recall from Chapter 5, the Sun's path changes throughout the year from a high point in June to the low point in December (Figure 6-5). The length of the overhang determines when the Sun's rays can enter a building.

## Passive Solar Design Options

Passive solar can be incorporated into almost any building. Even if the long axis is oriented north and south instead of east and west, windows can be placed in the south-facing walls to capture the low-angled winter sun. For the best year-round

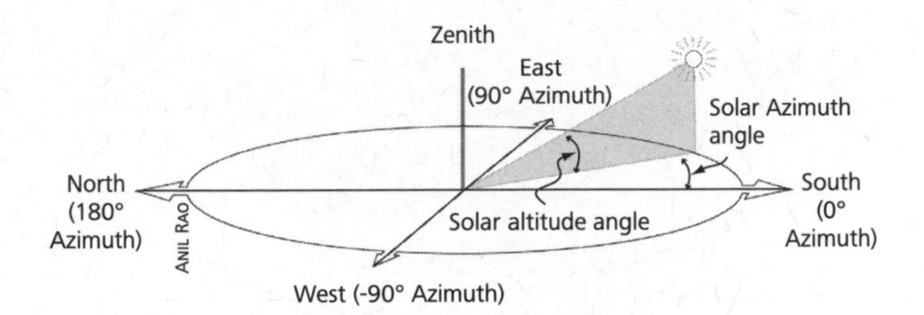


Fig. 6-5: *The Sun's position in the sky changes throughout the year, as illustrated here. It is lowest in the winter, which is ideally suited for passive solar heating.*

results, however, a passive solar building should be oriented to true south. Improper orientation, like orienting the long axis of a building from north to south will increase solar gain in the summer — often dramatically — which results in massive unwanted heat gain. This, in turn, results in considerable discomfort and high cooling costs. In addition, the wrong orientation dramatically reduces the potential for solar heating.

### *Direct Gain*

When it comes to designing a passive solar building, you have three options. The first is *direct gain*. As shown in Figure 6-6a, direct gain is the simplest of all passive solar design options. The building becomes a huge solar collector. South-facing windows on the building allow the Sun's rays to enter, heating the home directly, hence its name.

Direct gain is not only the simplest passive solar design option, it is also arguably the most efficient and cost effective, for reasons that will become clear shortly.

For best results, be sure to install thick, insulated window shades or curtains over all windows — especially the solar glazing — to hold heat in at night. At night, solar glazing can get quite cold. If it isn't covered, people sitting near the windows on cold winter nights will feel quite cold.

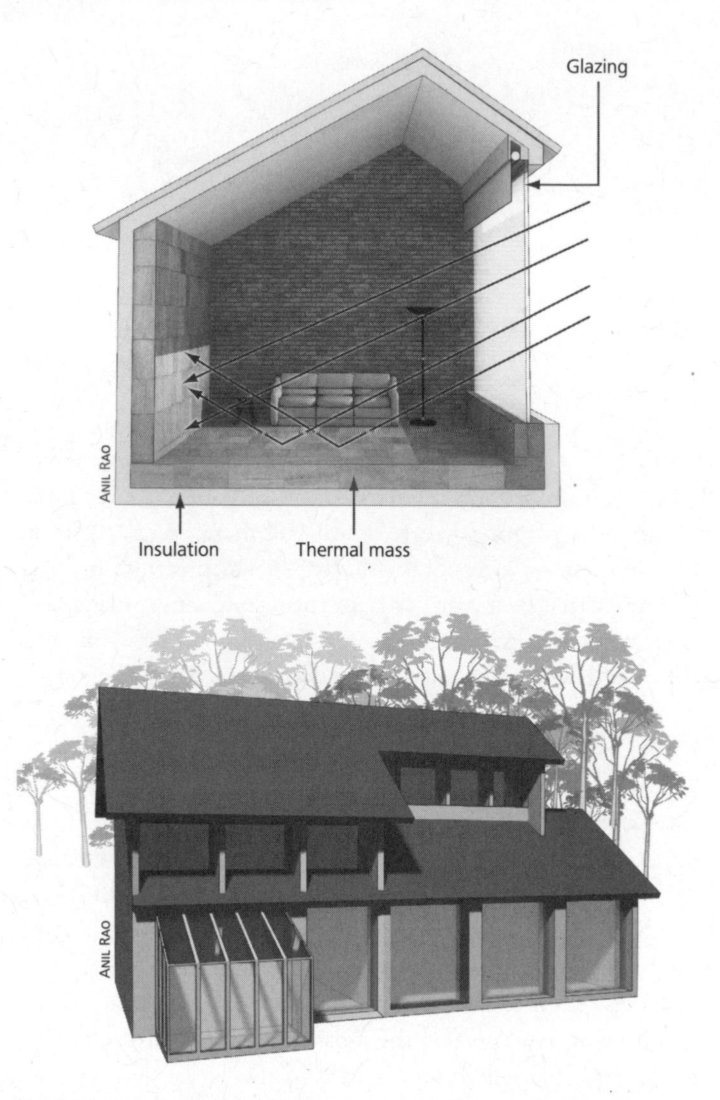


Fig. 6-6 a–b: *(a) In a direct gain passive solar building, sunlight enters the building through solar glazing, heating it directly. Thermal mass located in the direct path of incoming solar radiation is vital to achieving maximum comfort. Thermal mass can be located in floors and walls as shown here. (b) Isolated gain — attached sunspace.*

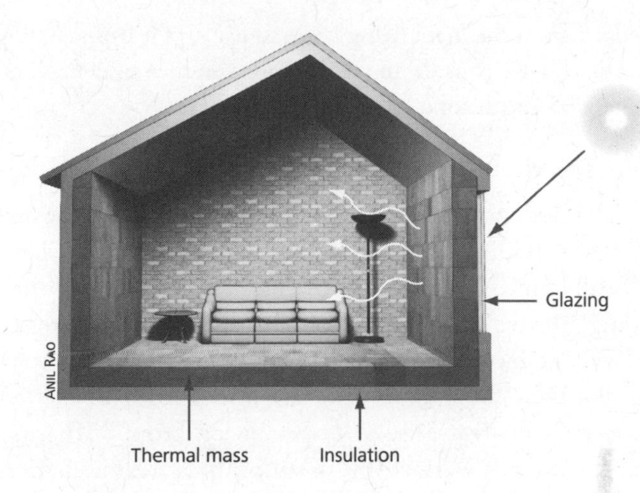


Fig. 6-6 c: *Indirect gain — Trombe wall.*

## *Isolated Gain — Attached Sunspaces*

Another option for passive solar gain is referred to as *isolated gain,* and is shown in Figure 6-6b. Isolated gain, also referred to as *attached sunspaces,* is great for retrofits (discussed in more depth in Chapter 7). Located on the south side of homes, attached sunspaces act as solar collectors. Solar-heated air warms the sunspace and, if it's designed properly, the adjoining rooms. In this design, then, sunlight is absorbed in a separate, or isolated, space, then transferred to the main living spaces — hence the name "isolated gain."

Attached sunspaces can be incorporated in new building designs, but they tend to increase the cost of a building. Direct gain is far cheaper when building anew. Attached sunspaces need to be designed very carefully to achieve one's goals. All-glass sunspaces, for instance, have a nasty habit of overheating, indeed baking, in the hot summer sun. They can even bake occupants on cold, but sunny, winter days, making the spaces virtually uninhabitable when the Sun's shining in. They also tend to get very cold at night in winter, so they need to be closed off from the main living space to prevent it from cooling down. If you are considering this option, be sure to consult an experienced and accomplished solar designer.

### *Indirect Gain — Trombe Walls*

The third option for passive solar is *indirect gain*. As shown in Figure 6-6c, indirect gain involves the installation of a large mass wall immediately behind windows on the south side of a building. The windows are typically two to six inches from the mass wall, no more.

Walls built for indirect gain are also known as *thermal storage* walls or *Trombe walls* (pronounced "trom"). Thermal storage walls are warmed by the incoming solar radiation. Heat generated in the walls migrates inward by conduction (moving from hot to cold) during the day. By the time the Sun sets, the heat has migrated to the interior of the wall. It then radiates into the adjoining rooms, warming them. Because the mass wall heats first, and because this heat is then transferred to adjoining rooms, this design is referred to as *indirect gain*.

Trombe walls are the invention of a French engineer by the same name. They work extremely well in all climate zones — if designed and constructed properly. (In many cases, though, they're poorly designed and fail miserably!)

Thermal storage walls provide delayed solar gain. That is, they heat up during the day, then transfer that heat to the adjoining rooms after the Sun sets. For immediate gain, windows can be installed in the thermal mass wall, allowing sunlight to penetrate directly into the room. In addition, vents can be installed in the upper and lower portions of the wall, as shown in Figure 6-7. The vents allow room air to circulate through the space between the thermal storage wall and the glass. Sunlight warms the air, causing it to expand. As air expands, it becomes lighter. The lighter, warm air rises, then pours into the room. This creates a thermosiphon, a force

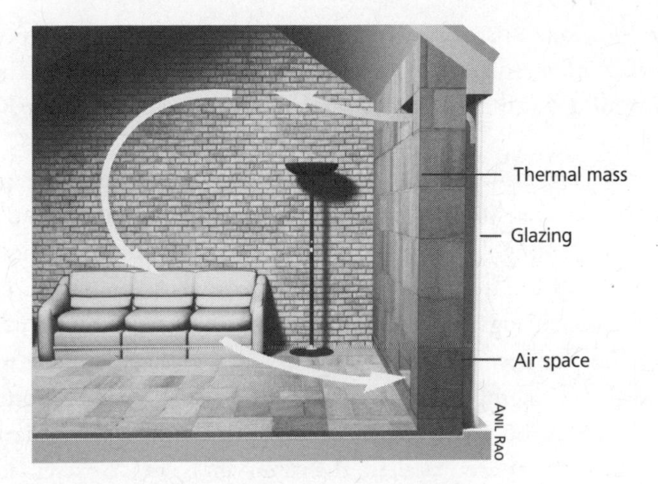


Fig. 6-7: *Heat given off by the thermal storage wall radiates into the room keeping it warm at night. A window placed in this wall would permits day lighting and daytime solar gain. Vents shown here also permit daytime heating.*

that pulls cooler floor-level air through the lower vent. This cooler air is then heated by solar radiation. As long as the Sun shines, this thermal convection loop continues to circulate air from the room to the airspace between window and Trombe wall where it is heated, then back into the room. Vents can be either manually or automatically shut at night to prevent reverse air movement at night, which would cool the room down. (At night, air between the mass wall and the glass cools and sinks, and could enter the room through open lower vents. This would draw warm room air through vents at the top of the wall, where it would be cooled.)

Indirect gain is the rarest of all passive solar options. It has a bad reputation in the solar industry, but my experience is that it works well — actually, it works extremely well — if it is designed correctly. I've inspected several homes in which it wasn't working, and in each case the homeowner had made an egregious mistake by installing sheet rock over the inside surface of the mass wall, screwing it onto furring strips. The layer of drywall blocks heat flow from the mass wall into the interior.

Thermal storage walls require heftier foundations to support their weight, so plan accordingly. If designed and built correctly, though, they should provide years of maintenance-free comfort. They're especially well suited to home offices occupied during daylight hours. The mass walls block out much of the incoming solar radiation, preventing overheating and glare on computer screens. They're also great in bedrooms for people who like to sleep in dark rooms at night. If dark window shades are installed, the room can be made extremely dark.

## Next Steps

Passive solar homes incorporate these design ideas that allow them to collect and store solar energy to heat homes during the day and night — and during cloudy periods. There is much more to passive solar design than I've covered here, but this provides the basics. I urge readers to study the topic in more detail either by attending workshops on passive solar design (like those I offer through The Evergreen Institute) or by reading more books and articles on the subject. Further study will help you better appreciate this design strategy. You may also want to take a look at my book, *The Solar House: Passive Heating and Cooling*. It is written for the general reader — you don't have to be an engineer to understand the material.

## Conclusion

Passive solar can be incorporated into any building design. No matter what your tastes are, you can easily incorporate passive solar, creating a home that is attractive and very economical to live in.

Passive solar heating is an intelligent choice in building design. It provides free heat — often an amazing amount of free heat — for the life of a building, which could potentially save owners hundreds of thousands of dollars in fuel bills. How much heat you can obtain from the Sun depends on the amount of sunshine available in your area and your attention to the principles of passive solar design, especially proper orientation, window placement, quality of windows, air sealing, and insulation. The more faithfully you follow the design guidelines, the greater your solar gain.

Even in the least sunny climates in the United States, it's possible to obtain 40% to 50% of your heat from the Sun. In sunnier climates, with careful design, 80% is a reasonable goal. Such efforts will not only help you save money, they'll help you dramatically reduce your carbon footprint.

Passive design also ensures a much cooler home in the summer. Proper orientation, window allocation, air sealing, insulation, overhang, and thermal mass — all key principles of passive solar design — also help keep a home much cooler during the dog days of summer. Dollars or Euros invested in passive solar therefore pay double dividends.

It typically costs nothing more, or at best only a tiny bit more, to build a superefficient passive solar home that stays warm in the winter and cool in the summer. The question isn't "Why should you build a passive solar home?" but "Why wouldn't you?"

CHAPTER 7

# RETROFITTING YOUR HOME FOR PASSIVE SOLAR

If building a brand new passive solar home isn't in the cards for you, don't despair. It is often possible to retrofit an existing home to incorporate this amazingly clean, low-tech, cost-effective, and environmentally friendly way of heating. This chapter tells you how. As you shall soon see, there are three main options: opening up the south facade with windows that allow the low-angled winter sun to enter; incorporating passive solar into an addition; or adding an attached sunspace, a topic briefly mentioned in Chapter 6.

## Option 1: Installing South-Facing Windows

To add windows to an existing home, your house must have one outside south-facing wall with unfettered access to the winter sun. This is the wall in which the additional windows will be installed. No trees or buildings can shade the wall during daylight hours — ideally from 9 am to 4 pm — if you want the retrofit to perform well. And the south-facing wall must be in line with incoming sunlight throughout most of the day throughout the heating season. In many parts of the world, the heating season starts in mid-to-late fall. It extends through the winter, and ends in mid spring. In colder climates, the heating season may begin very early in the fall and extend to late spring.

Newly built passive solar homes are typically elongated structures, often rectangular in shape. The rectangular footprint is oriented so that the long axis of the building runs from east to west. This allows the designer to "load up" the south-facing facade with windows to permit maximum solar gain.

In a passive solar retrofit, the south-facing wall doesn't have to run the length of the building, although that's ideal. Basically, the more south-facing wall space, the better.

In North America, many houses are oriented north–south. That is, their long axes run north and south, so they face the street. Because most are squeezed in on narrow lots, their south-facing exterior walls are pretty small. Don't despair. Adding just a few windows can boost your solar gain and provide heat. You won't be able to heat your entire home, but you can provide heat to a couple of rooms, helping to reduce your energy bill. Before you start installing windows, however, there are a couple things you need to do.

### *Step 1: Seal up the Leaks*

Passive solar retrofitting begins not by adding windows, though this is an important measure, but by air sealing and insulating a building to the max. Collectively, this is referred to as weatherization, and is discussed in Chapters 3 and 4. For best results, be sure to conduct an energy audit or hire a professional to do the job to determine where leaks are and where and how much insulation must be added. If you haven't read those chapters, be sure to do so.

### *Step 2: Insulate, Insulate, Insulate!*

After the leaks have been sealed, it is time to insulate, based on your findings or the findings of the professional auditor. Be sure to insulate the walls, ceilings, and the foundation, as outlined in Chapter 4. Also be sure to insulate under floors over unconditioned spaces such as unheated basements and crawl spaces. It's also a good idea to install insulated window shades over all windows to hold heat in at night. And don't forget to use them!

Sealing a home or office, then insulating it, can reduce heating (and cooling) costs dramatically. Weatherization also helps your home retain the solar heat you invite into the building by installing windows along the south facade.

### *Step 3: Installing Energy-Efficient Windows*

After you have sealed and insulated your home, it's time to install windows on the sunny south side of the building to increase its direct solar gain — assuming, of course, you have the space for windows.

Simple as it may sound, adding a few windows on the south side of a home is no easy task. It requires a lot of knowledge and skill and can be tricky. You'll probably need a building permit. Because the task is so complex, it's a job usually best left to building professionals — either skilled handymen or handywomen or professional window installers.

A new energy-efficient window is only as good as its installation. If not done right, the window can leak, allowing tons of cold air in. Improper installation can trigger a cascade of problems. In addition to damage to the window itself, improperly installed windows can let moisture penetrate between the panes of glass. It can also allow water to leak into your walls, which can reduce the R-value of your insulation, promote mold growth, and even cause structural damage.

To properly install a window in an existing wall, you or the installer will first need to cut a hole through the siding, exterior sheathing, and interior wall (typically covered with drywall, plaster, or paneling). Once the hole is cut, the installer removes the existing studs (wood framing members) and installs a rough window opening (basically, a box to which the new window will be attached). This requires a header — a horizontal framing member over the top of the rough window opening, as shown in Figure 7-1. The header supports the weight of the wall above the window and transfers the weight to the framing members along the sides of the windows. This diverts pressure that would otherwise press directly onto a window, causing it to crack. Another way of saying this is that the headers prevent the weight of wall and roof above windows from pressing down on them.

After rough window openings have been created and properly framed in, it is time to install the solar windows. The windows you install must be sized about a half an inch smaller than the rough opening. This allows room for expansion in hot weather. Windows come in wooden, vinyl, fiberglass, or metal frames that attach to the rough opening. Be sure to read and follow directions provided by the manufacturer very carefully.

Windows are attached on the outside to the sheathing by a metal or plastic flange that attaches to the window frame.

DAN CHIRAS

Fig. 7-1: *This opening in my office was created to install south-facing windows to permit passive solar gain. Note that the vertical framing members (studs) have been removed and a header has been installed along the top of the opening.*

Nails are driven into the flange to hold the window in place. Be sure the window is level before you nail it in place.

The gap between the window and the frame is then filled from the inside with foam insulation such as backer rod or the type of spray foam insulation designed specifically for windows and doors (Figure 7-2.) Using the wrong product, for example, the spray foam used for sealing large gaps and cracks, can void the window manufacturer's warranty, so read instructions carefully — or be sure your installer knows what he or she is doing.

After the window is secured, be sure to apply adhesive sill flashing. It seals the window opening from the outside, as shown in Figure 7-3. This prevents air and moisture from penetrating from the outside.

DAN CHIRAS

DAN CHIRAS

Fig. 7-2: *This product (made by Dow) does not expand as much as other liquid spray-in foam sealants, which can cause window frames to warp. It is therefore ideal for window and door installations. Some manufacturers require installers to use foam weather stripping instead to ensure that there is no expansion and so no damage to the newly installed window.*

Fig. 7-3: *Newly installed windows should be sealed with this adhesive tape to prevent air and moisture from penetrating.*

Once the window is in place, you or the installer will need to install trim. Be sure to caulk around the trim to ensure an airtight seal.

## *What Type of Window?*

As you shop for windows, you will find a wide range of types and an equally wide range of prices. When it comes to windows, cheap is rarely better. Like so many things in life, you get what you pay for. A high-quality window will be more expensive, and it will generally be better built, more energy efficient, and more durable than a less expensive window. So, when it comes to purchasing windows, buy the best ones you can afford.

The most airtight windows on the market are those that don't open — inoperable windows. These are ideal for solar glazing (south-facing glass) in passive solar retrofits, but there has to be at least one operable window in a room. (Building codes require an operable window for egress — escape in case of a fire — and you'll need one for natural ventilation, too.)

If you want to or need to install operable windows, your best bet is to purchase hinged windows — that is casement, hopper, or awning windows. These windows open with the aid of a crank mechanism and shut very tightly against weather stripping — provided the window is well made.

Slider windows — single-hung, double-hung, and horizontal — tend to be less airtight and aren't usually a good choice — unless you buy really high-quality ones. Don't forget, quality is everything. A high-quality slider, in fact, could be more airtight than a cheaply made hinged window.

Solar glazing requires special windows designed to permit a lot of sunlight to enter. Be sure to ask for windows with a high solar heat gain coefficient (SHGC). Although SHGC sounds like a complicated engineering term, it's actually quite simple. SHGC is a number from 0 to 1 that indicates the amount of solar energy that passes through a window. A window opening covered with plywood allows no solar energy to enter. It is, therefore, assigned a SHGC of 0. A window opening with no glass would have a SHGC of 1, indicating that it allows 100% of the incoming solar energy to enter.

For solar glazing (south-facing windows), the higher the SHGC, generally, the better. If you are in a cold climate, buy windows with SHGC of 0.5 or higher. In slightly warmer climates, windows with a lower SHGC will work, from 0.4 to 0.5.

Buying high-SHGC windows can be a huge challenge because they're not typically stocked. Contact local window suppliers to see if they can order them for you. Window manufacturers can buy high-SHGC glass from PPG, a company based in Pittsburgh, PA. They offer two products: Sungate 100 and Sungate 500, both of which are high-SHGC glass. Window manufacturers can order and then install the glass in their own frames. (Individuals and businesses can't buy directly from PPG. They are strictly a glass manufacturer and only sell glass to window manufacturers.) Thermotech, in Canada, also sells windows containing high-SHGC glass.

South-facing windows live a rough life. They are exposed to lots of sunlight, so buy high-quality windows. I recommend wood-framed windows with an exterior cladding of aluminum to protect against the weather. Fiberglass and vinyl windows work well, too, but they're not as environmentally friendly as wood-framed windows. Stay away from aluminum-framed windows unless the frames are very well insulated. Otherwise, the metal frames will lose a lot of heat in the winter by conduction. Also, be sure to buy windows with warm-edge spacers. Spacers are foam pieces that are inserted between the two panes of glass, along their perimeter, blocking heat flow through the edges. In other words, they reduce heat loss around the perimeter of a window.

To learn more about windows, you can check out one of my books, either *Green Home Improvement* or *The Solar House*. Both have lengthy discussions of windows. I also strongly recommend that you contact a number of window installers in your area. Explain your goals — to increase solar gain — and ask them what products they have that will allow you to capture the low-angled winter sun. Be sure they understand your needs and can deliver windows with the proper SHGC.

When retrofitting a home for passive solar, it's a good idea to reduce the number of windows on the east, west, and north sides. North-facing windows are especially troublesome, as they tend to lose huge amounts of heat on cold winter days. East- and west-facing windows can allow a lot of solar gain in the summer, leading to overheating. Unfortunately, removing windows is usually a costly venture, so it may not be feasible. It may also have an adverse effect on the appearance of your home, so may not be desirable. As noted earlier, though, be sure to install insulated window shades or shutters over all windows — especially the new windows — to decrease nighttime heat loss.

### *Adding Thermal Mass*

If you end up dramatically increasing the amount of south-facing glass, you may want or need to add thermal mass. As you may recall from Chapter 6, thermal mass is any masonry material that can absorb heat generated by incoming solar energy during the day and release it at night or on cold, cloudy days. Thermal mass serves as a heat sponge that prevents overheating during the day. At night, or on cloudy days, thermal mass releases stored heat. As a result, thermal mass helps maintain a more consistent internal temperature in a passive solar home.

Thermal mass is easiest to install in new construction; it is much more challenging to install in retrofits. Floor tile or brick facing added to nearby walls may help. Adding a second layer of drywall to existing walls in direct contact with the incoming solar radiation works well, too. For best results, the thermal mass must be directly in the path of incoming sunlight. Mass should be a darker color to increase the absorption of sunlight. You don't need to paint your walls black or dark brown. However, a dark green wall can help improve the performance of drywall mass.

One note of caution: For best results, the south-facing windows you are adding should be shaded by eaves (overhang) to prevent heat gain during the cooling season — late spring, summer, and early fall. You may need to build eaves over newly added solar glazing if there is no overhang on the south side of the building. You can also install retractable awnings that can be opened in the summer to prevent overheating.

Adding new windows is not cheap by any stretch of the imagination, but it will help improve the energy performance and comfort of your home, if done correctly. Reducing heating costs obviously saves on fuel bills. Adding solar glazing will increase the amount of day lighting — natural lighting during daytime hours. This, in turn, will reduce electrical costs by reducing the amount of artificial lighting you'll use.

## Option 2. Building a Solar Addition

If you are planning on increasing the square footage of your home, you may want to consider designing it for passive solar — provided at least one wall of the addition will be unshaded during the heating season. Installing solar glazing on the south-facing wall of an addition could, if done correctly, provide up to 80% of the heat required by the addition, perhaps even more in warmer climates. A solar addition can also provide a little heat to an adjoining room, if done correctly (Figure 7-4).

Solar additions can be designed as either direct gain or indirect gain structures, both described in the previous chapter. The least expensive is direct gain. In direct gain designs,

DAN CHIRAS

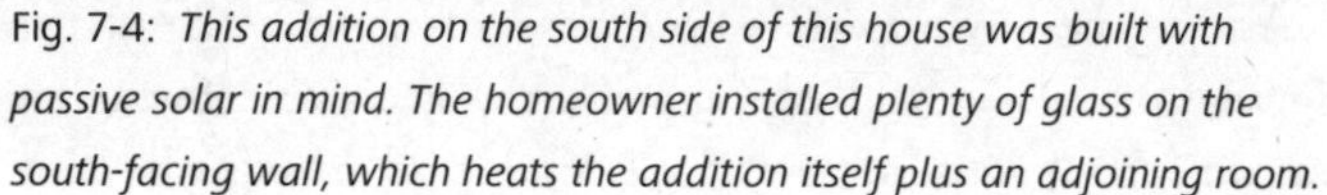

Fig. 7-4: *This addition on the south side of this house was built with passive solar in mind. The homeowner installed plenty of glass on the south-facing wall, which heats the addition itself plus an adjoining room.*

additions serve as huge solar collectors — just like any other direct gain building. Solar energy heats the addition but as a rule, very little heat is available to adjoining rooms. Indirect gain (incorporating a thermal storage wall) works well for bedroom and office additions, but it costs more than direct gain, because it requires the installation of thermal mass. Additional thermal mass also requires a sturdier foundation, which costs more. However, the benefits could easily outweigh the higher costs.

When designing a solar addition, be sure to follow all the rules I outlined in the last chapter. Orient the addition to south (long axis east and west), concentrate the windows on the south, install adequate overhang to shade the structure in the summer, make it super airtight, and insulate it extremely well. I advise installing thermal mass if the solar glazing exceeds 7% of the floor space of a single-story addition.

Be sure to install energy-efficient, high solar heat gain coefficient glass on the south wall. Be certain not to overglaze, however, or the room may overheat. (Generally that means not exceeding 15% solar glazing.) If you want the structure to provide heat to neighboring rooms, you will need to find a creative way to move heat from the addition to adjoining living areas. Because this can get quite tricky, it is a good idea to hire an architect, an engineer, or a builder who is experienced in passive solar. An experienced passive solar consultant could help as well. Experienced professionals can help you work out the details so the addition will perform well — heating itself naturally without baking you and your pets in the process.

## Option 3: Attached Sunspaces

Attached sunspaces — a.k.a. solar greenhouses — are often the easiest way to incorporate passive solar in an existing building. This design, known as *isolated gain,* is simple in theory, but very difficult to get right. So, proceed carefully.

Be wary of all-glass attached sunspaces. They are attractive, fairly inexpensive, available in kits, and required for growing vegetables year round (Figure 7-5). They're often touted as three- or four-season rooms.

Unfortunately, all-glass attached sunspaces tend to overheat in the summer and can overheat during the heating season — even in the dead of winter. This often makes them virtually uninhabitable during daylight hours on many days. Intense sunlight is very rough on furniture and carpets. It's even harder on tropical house plants. As you may know, many tropical plants grow in the shade or in less-intense sunlight than they will encounter in an all-glass greenhouse, especially in structures at higher elevations like Colorado where sunlight is more intense than at lower-elevations (there's less atmospheric filtering of incoming solar radiation in high-altitude areas). During the summer, intense heat is very hard

DAN CHIRAS

Fig. 7-5: *This attractive solar collector built on the south side of a house may seem like a good idea, but it suffers serious problems, like overheating in the summer. These shades must be drawn to protect the space from overheating on sunny days throughout the late spring, summer, and early fall.*

on vegetables. Most garden vegetables do very poorly in an attached sunspace if temperatures skyrocket, which they often do. Photosynthesis starts to slow down at 85° F and virtually stops at 100° F.

While all-glass attached sunspaces may generate a lot of heat, it's not that easy to force heat to flow laterally to an adjoining room. Hot air rises, but it doesn't move sideways very well unless forced to do so. That means you'll need to devise some system to move heat out of the sunspace. A thermostatically controlled wall fan is often a good solution. It can blow heat from the attached sunspace into neighboring rooms.

As just noted, all-glass attached sunspaces tend to be drenched in sunlight during the day and are therefore too hot and sunny to occupy. At night, they tend to get very cold, making them pretty useless as additional living space. Because they can become extremely cold at night, attached sunspaces should be closed off from adjoining rooms to keep them from chilling down.

A far better option is an attached sunspace with south-facing glass and a solid roof, much like a solar addition (Figure 7-4). An attached sunspace with a roof will gain less heat during the day than its all-glass cousin, but will be more comfortable year round. Unfortunately, this arrangement isn't well suited for growing plants in the summer. Because the Sun is high in the sky, plants will receive very little sunlight. As a result, they tend to grow spindly, or leggy, without overhead lighting. When designing an attached sunspace of this type, be sure to follow all of the rules for direct gain solar structures.

The amount of heat an attached sunspace delivers to the rest of the house depends on several factors. One of the most important is the amount of glass installed on the south-facing wall. The more glass you install, the more heat the sunspace will generate.

Also important is the amount of thermal mass in the attached sunspace. The more thermal mass that's installed, the more heat that will be absorbed and used within the structure itself. When lots of thermal mass is installed, you create a space that heats itself, but has very little, if any, heat to give to adjoining rooms.

If you want an attached sunspace to heat adjacent rooms, you will need to install a lot less thermal mass in the sunspace — perhaps no additional thermal mass at all. This strategy turns the attached sunspace into a giant solar collector. The sunspace will, however, probably become unbearably hot during the day, unless you force hot air into neighboring rooms. A quiet, energy-efficient, thermostatically controlled fan mounted in the wall of the sunspace can move air into neighboring rooms. Another approach I recommend in many solar designs I consult on is to install a thermostatically controlled fan in the ceiling of the sunspace. It is attached to a duct that blows the hot air into adjacent rooms. (The duct runs through or below the ceiling into neighboring rooms.) Be sure to seal and insulate ductwork and keep duct runs as short as possible. Also be sure to install a quiet (low sone) fan.

Whatever attached sunspace style you choose, you should consider designing it so you can close it off from the main living area at night. This thermally isolates it from the rest of the home at night or on cloudy days, preventing heat from the house from moving into the sunspace and then into outer space, where it does you no good. You should also consider installing insulated shades to prevent heat loss at night, especially in an attached sunspace you are hoping will provide additional living space. For more details on attached sunspaces, you may want to check out my book, *The Solar House: Passive Heating and Cooling.*

## Conclusion

Adding solar windows, creating a solar addition, or building an attached sunspace on the south side of a building can increase your reliance on clean, abundant, renewable, and economical solar energy. It will cut your fuel bill and add to the value of your home. As the price of fuels rises, and as concern over climate change increases, energy-efficient solar homes are likely to be hot sellers in the real estate market. No matter what you do, though, proceed with caution. Be careful. Pay close attention to air sealing and insulation. Seek professional advice. If you design and build correctly, you will be granted many years of free heat.

Chapter 8

# Solar Hot Air

Another option for clean, affordable heat is a solar hot air system. Solar hot air systems capture sunlight energy striking a collector mounted on or near a building. The solar hot air collector heats indoor air that is circulated through the collector, then sends it back into the building. These systems can provide years of inexpensive, worry-free comfort.

Solar hot air systems are primarily installed on houses, but they can also be used to heat office buildings, workshops, garages, and barns. Very large systems can be installed on warehouses, factories, and other big commercial buildings.

Like other renewable energy technologies, solar hot air systems free homeowners from worries over rising fuel costs. Solar energy costs the same today as it did when humans first appeared on the savannahs of Africa — nothing. In this chapter, you will learn how solar hot air systems work, how they are installed, and what the costs and return on your investment might be.

## How Do they Work?

Solar hot air systems are fairly simple devices, far simpler and easier to install than passive solar or solar hot water space-heating systems, the topic of Chapter 9. They're also a lot cheaper, and do not require complicated electronic controls. If you're skilled, you can even build a solar hot air system yourself (Figure 8-1).

Solar hot air systems produce heat earlier and later in the day than solar hot water systems. As a result, "they may produce more usable energy over a heating season than a liquid system of the same size," according to the DOE's online publication, "Consumer Guide to Energy Efficiency and Renewable Energy." Moreover, air systems do not freeze. Minor leaks in the collector or distribution ducts that can cause significant problems in hot water systems are of lesser consequence in hot air systems. So what does a solar hot air system look like?

Fig. 8-1: *It's not that difficult to build a solar hot air collector. (a) This solar hot air panel designed and built by Missouri resident Mike Haney shows what's possible with a little time and a lot of skill. (b) This fairly crude but effective system is used to heat a house in Colorado Springs, Colorado.*

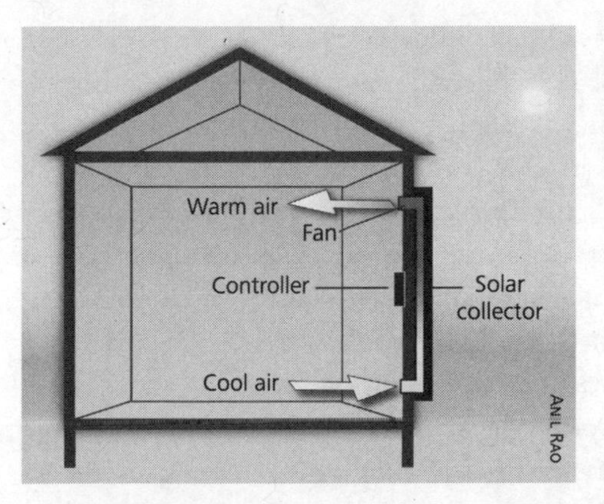


Fig. 8-2: *This diagram shows how air is circulated through and heated in a solar hot air collector.*

All solar hot air systems rely on a collector (described shortly). The collector is typically mounted on the roof of a house, but may also be mounted on vertical south-facing walls or even on the ground (if it's not shaded during the heating season). As you shall soon see, vertical south-facing walls are the best location.

Solar hot air collectors capture sunlight energy and use that energy to heat room air that circulates through them (Figure 8-2). The solar-heated air is transferred into the interior of the building, thanks to a small AC or DC fan or blower.

Solar hot air systems are controlled by relatively simple electronics (unlike solar hot water systems). A temperature sensor mounted inside the panel monitors collector temperature. When it climbs to 110° F, the sensor sends a signal to a thermostat inside the home. It, in turn, sends a signal to the fan or blower, turning it on when room temperature drops below a desired temperature setting. When the temperature inside the collector drops to 90° F, the fan switches off.

Solar hot air systems provide daytime heat on cold, sunny days (unlike solar hot water systems, which are designed for daytime and nighttime heat thanks to their ability to store heat in water tanks). Some early systems stored hot air in rock storage bins for use at night or during cloudy periods. Although ingenious, this approach generally proved to be disappointing — quite disappointing. Designers found it difficult to achieve adequate and predictable airflow through rock beds. Rock storage also poses a health risk. As green building expert Alex Wilson notes, rock beds "can become an incubator for mold and other biological contaminants — causing indoor air quality problems in the home." Because of these problems, most rock bed storage systems have been abandoned.

Even though solar hot air systems used today principally provide daytime heat, the heat they produce can accumulate indoors during the day. For example, it may be absorbed by drywall, tile, or the framing lumber of a building. In the early evening, the heat stored in these forms of thermal mass radiates into the rooms, helping create evening comfort. Obviously, the more thermal mass, the greater the nighttime benefit. Even so, solar hot air systems are still primarily considered daytime heat sources.

## Types of Solar Hot Air Collectors

Although you may not have heard about solar hot air systems, they're not a new technology by any means. They've been around since the 1950s. Today's systems fall into two categories: open- and closed-loop.

### *Open-Loop Solar Hot Air Collectors*

The newest solar hot air system is the open-loop design (Figure 8-3). Hot air collectors in open-loop systems extract cold air not from the building, but from the out-of-doors. They heat this air, and then transfer the heated air into the building. Collectors used in open-loop systems are known as *transpired air collectors.*

A transpired air collector consists of a dark-colored, perforated metal facing, known as the *absorber plate* (Figure 8-4).

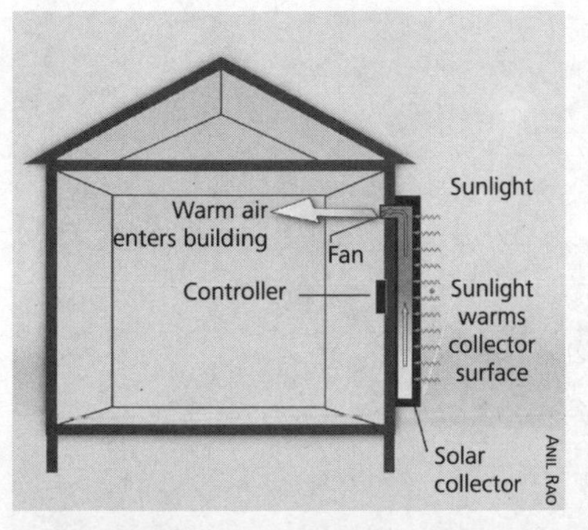


Fig. 8-3: *In this design, cold outside air is drawn into the collector where it is heated. The warmed air is then blown into the building. These systems are primarily used in large warehouses, which are not very airtight and need only a modest amount of heat.*

Fig. 8-4: *Absorber Plate of Solar Hot Air Collector.*

The sides and back of the solar hot air collector are made from metal and are typically insulated to reduce heat loss.

Sunlight striking the absorber plate of a transpired collector heats the surface. Heated air is drawn into the collector by a blower and is then piped to the interior of the building.

## *Closed-Loop Solar Hot Air Collectors*

A closed-loop solar hot air system, the most popular and effective option, consists of a glazed flat-plate collector (Figure

DAN CHIRAS

Fig. 8-5: *This collector from Your Solar Home is mounted on my garage. It contains its own source of electricity, a small solar panel that powers the fan that moves air through the collector.*

8-5). Inside the collector is an absorber plate. (The absorber plate has a rough surface that increases turbulence inside the device. Turbulence helps strip heat from the absorber plate). The collector is insulated and covered with single- or double-pane glass (glazing).

In closed-loop systems, cool air is drawn into the collector from the interior of the building and then heated as it passes through the collector. The heated air is then blown into the building. Room air is typically drawn into the panel through a short section of pipe that runs through the wall or the roof of a building, although longer duct runs are used in some applications.

Air entering the intake duct passes through the collector, either behind the dark-colored metal collector surface (back-pass collectors) or in front of it (front-pass collectors). Back-pass models come with single-pane glass; front-pass models require two-pane glass to reduce heat loss. Heated air is blown into the building through a pipe or duct (kept as short as possible). Small registers are mounted on the air intake and outtake openings that penetrate the exterior walls of the building.

Of the two types of glazed solar hot air collectors, back-pass collectors are most common. They're cheaper to manufacture and about 50 pounds lighter than front-pass collectors and, therefore, are a little easier to install.

Like transpired air collectors, glazed collectors in closed-loop systems are thermostatically controlled. Both closed-loop and open-loop systems employ backdraft dampers to prevent air from flowing in reverse at night via convection, a natural phenomenon that will suck heat out of a building.

## Installing a Solar Hot Air System

Installing a solar hot air collector is not a job for novices. Although, as solar expert Chuck Marken notes in his article on solar hot air systems in *Home Power* magazine, "a seasoned crew of two can install a solar hot air system in a few hours...[but] if this is your first time, plan on a weekend, even with help."

### *Mounting Options*

As noted previously, the best place to mount a solar hot air collector is on the south side of a house, provided it is unshaded during the heating season. Mounting the collector vertically on a south-facing wall ensures maximum solar gain during the coldest months — when you need the heat the most (Figure 8-6). (Vertical mounting means the collector is more closely aligned to the incoming solar radiation from the low-angled winter sun.) Mounting on a south-facing wall also shades the solar hot air collector during the summer from the high-angled summer sun, especially if the building has adequate overhang.

The second most desirable location for a solar hot air collector (though infrequently chosen, according to Todd

YOUR SOLAR HOME

Fig. 8-6: *Mounting a solar hot air collector on the south side of a building is ideal.*

Kirkpatrick of Your Solar Home) is on a rack mounted on the ground. This is known as a *ground-mounted solar hot air collector.* The rack must be anchored to a suitable concrete foundation, for example, concrete piers.

Ground-mounted systems require considerable ductwork — as do roof-mounted systems, discussed shortly. Due to the large amount of ductwork, both systems require sizable fans to ensure adequate airflow.

Because they function primarily during the heating season — when the Sun is low in the sky — solar hot air collectors should be mounted more vertically than solar electric modules, which are mounted to absorb as much sunlight as possible through the whole year. The *tilt angle* of the solar hot air collector, that is, the angle between the back of the module and a line running horizontally from the bottom of the collector, should be set at latitude plus 15°. If you live at 45° north latitude, for instance, you should mount a collector at 60°. If you live at 35° north latitude, the tilt angle should be 50°.

One of the most popular of all places to mount a solar hot air collector, though one of the most problematic, is on the roof. Roofs are popular because they often are unshaded during the heating season. And because roofs often tower above other buildings — and even mature trees in some cases — they provide good access to the winter sun in densely populated urban and suburban environments.

Unfortunately, roof mounts can be more complicated and more costly than wall or ground mounts. In homes with attics, for instance, installation requires the use of flexible insulated ducts to transport air to and from the collector. Unfortunately, flexible insulated ducts are ribbed, which greatly increases air turbulence, which reduces airflow. Partially because of this problem, much larger fans are required in roof-mounted systems. In homes with closed ceiling cavities, collectors require much shorter duct runs however. The same goes for wall-mounted collectors.

Another problem with roof mounts is that they are exposed to sunlight year round, while most of us need heat only during the late fall, winter, and early spring. Intense sunlight during the summer can, over the years, damage a roof-mounted solar hot air collector, so professionals generally recommend against such installations. However, if that's the only location you have, you can still install one, but cover the collector during the summer.

## *Pointers for Mounting a Collector*

Precise directions for installing a solar hot air collector are beyond the scope of this book, however, a few comments and suggestions are in order — so you know what you are getting into.

To install a glazed solar hot air collector, you'll need to cut two large (five to seven inch) holes in the wall or roof and ceiling. When cutting holes in a wall or roof, be certain not to damage water pipes or cut or damage electrical wires. Work slowly, checking for potential obstructions. Cut a hole in the drywall or siding, then check for wires and pipes. Manufacturers provide metal mounts to attach to the exterior wall (or roof). The solar hot air collectors are attached to these mounts.

Because collectors are heavy and large, measuring around four by seven feet, you'll need a couple of brawny assistants to make sure the job goes right and to protect your back and toes. (You don't want to drop one of these collectors on your foot!)

When installing a collector, you'll also need to do some wiring, for example, you'll need to run electric wire to energize the fan or blower. You will also need to wire the thermostat (it comes with the unit) to the temperature sensor mounted in the collector. Wiring diagrams can be difficult to understand for the electrically illiterate.

To make your life easier, two manufacturers have provided some rather ingenious and simple wiring alternatives. Canada's Your Solar Home, for instance, manufactures a solar hot air collector known as the SolarSheat (Figure 8-5). It comes with its own supply of electricity: a small solar electric module mounted above the solar hot air collector. The solar electric module generates direct current (DC) electricity when struck by sunlight. The DC electricity it generates powers SolarSheat's DC fan. All the installer or homeowner needs to do is to wire the system to two wires that connect the collector to the thermostat inside the house. It's about as simple as you can get, as I learned when I installed mine.

Another ingenious solution to simplify wiring is provided by Cansolair. Cansolair's Solar Max is a glazed solar hot air collector made from 240 empty aluminum cans (Figure 8-7). The cans are painted black, arranged in 15 vertical columns, and housed in an attractive collector. Air flows through the solar-heated cans inside the collector. Airflow is supplied by an indoor fan that plugs into a 120-volt wall outlet. It propels room air through the duct system and then through the collector, and back again into the building. The fan, which is housed in an attractive console, also contains a washable filter that helps remove large particles from the air (Figure 8-8).

Although solar hot air collectors are often attached to existing walls or roofs, DeSoto Solar sells collectors that can be integrated into walls, reducing the collector's profile. Your Solar Home also has collectors that can be integrated in the exterior walls, but these products are best for brand-new construction or additions. It is much easier to install a collector in a wall before the sheathing and siding have been installed.

CANSOLAIR

Fig. 8-7: *Cansolair Solar Hot Air Collector.*

CANSOLAIR

Fig 8-8: *The Solar Max fan and filter conveniently plug into a wall socket, eliminating complicated wiring.*

### *How Well do They Work?*

Solar hot air systems boost the temperature of air flowing through them — often quite substantially. According to the DOE, "Air entering a (glazed) collector at 70° F (21° C) is typically warmed an additional 70–90° (21–32° C)."

Transpired air collectors may provide considerably less heat than glazed collectors. Solar energy expert Chuck Marken notes that transpired air collectors only increase the temperature of the air flowing through them around 11° F (52° C), which is of little value for residential structures, although it can be useful in factories, warehouses, indoor lumber yards, livestock enclosures, and the like, where a little bit of warming can dramatically improve working conditions for employees.

Ralf Seip, a homeowner who installed a SunAire glazed solar hot air collector to heat his basement workshop in Michigan, found that his collectors raised the temperature of the air flowing through them slightly less than 70° F (21° C), but only for an hour and a half to two hours a day during the peak of solar gain on sunny winter days. (That is, when solar irradiance is at its highest.) During the rest of the day, the system elevated temperature, but not as much as during peak sun. His system raised room temperature in his basement workshop by about 3.6° F (2° C) on cloudy days and 11° F (6° C) on sunny days. Although daytime temperatures in the workshop only reached 63° F (17° C), that was suitable for working.

Individuals who've installed solar hot air systems report impressive results. Steve Andrews, a residential energy expert based in Colorado, for example, installed a collector to heat the bottom 500 square feet of his tri-level home in sunny Denver. This area was usually 5–6° F colder than the rest of the house. He found that the solar hot air collector "made a difference during sunny winter days and the following evenings." Although the system was of little or no help on very cloudy or snowy days, "overall, the comfort improvement was dramatic."

The SolarSheat 1500G I tested while researching an article on solar hot air collectors for *Mother Earth News* (from which this chapter was adapted), consistently raised the temperature of indoor air entering the collector at around 68° F (20° C) by 40° F (4.4° C) on sunny cold winter days. It didn't make much difference in the room temperature, but I was dumping the heat into a very large open space of over 2,400 square feet. I'm certain the collector would have made a substantial contribution in a smaller room. This leads us to an important question: How much space can a solar hot air collector heat?

Solar expert Chuck Marken recommends one 4 x 8 foot collector per 500 to 1,000 square feet of heated space, depending on the solar resources at one's location and the energy efficiency of the building. A 2,000-square-foot home, for instance, would require two to four collectors. Separate collectors may be required for each room. For larger rooms, it may even be necessary to install two or more collectors, along with a more powerful fan to ensure adequate airflow. If collectors are shaded during part of the day (by tree limbs, for example), more collectors would be needed.

Remember! Whatever you do, be sure to seal up the leaks in your home and beef up the insulation *first*.

## Does Solar Hot Air Make Cents?

Bill Hurrle of Bay Area Home Performance in Wisconsin notes that the best his company's solar hot air systems can achieve (in their cold, cloudy climate) is a 25% to 35% annual reduction in heat bills.

Although this may not sound impressive, that reduction could cut fuel bills by $200 to $270 per year. And greater savings can be achieved in sunnier climates. As a rule, active solar heating systems are most cost effective in cold climates with good solar resources. But they also make sense elsewhere. A good solar installer will help you determine if a solar hot air system makes sense. Or, you can run the numbers yourself.

One popular way of determining the cost-effectiveness of a solar system is to calculate the return on investment. Return on investment is determined by dividing the annual savings by the cost of the system. If a $2,100 system saves $300 per year in heating bills, the return on investment is $300/$2100 or 14.3%. Manufacturers estimate returns on investment of 12.5% to 25% (based on current energy prices).

To accurately calculate return on investment, you should take into account the rising cost of fuel and any interest you pay on money borrowed to purchase the system, or lost interest if you withdraw the money from a savings account.

To get an even more precise number, maintenance costs should also be added to calculations of return on investment. Fortunately, very little maintenance is required on a solar hot air system; they only have two moving parts: a fan or blower, and a backdraft damper. Fans or blowers may need repair or replacement, but not for 18–25 years. Backdraft dampers can also be counted on for many years of trouble-free service. The rest of the system should last 50 years or longer!

When calculating return on investment, however, don't forget to check into financial incentives from state and local governments and local utilities. At this time, no federal incentives are offered for solar hot air systems unless they are dual-use systems — that is, systems that use some of the hot air to heat water for domestic uses. To avail yourself of these credits, however, the solar hot air collector must be SRCC (Solar Rating and Certification Corporation) certified.

## Shopper's Guide

Solar hot air collectors can be ordered online or purchased through a growing list of solar suppliers — companies that also install other solar systems such as solar hot water or solar electric systems. But shopper beware! "Marketing departments can make anything look good," says Bill Hurrle. Watch out for too-good-to-be-true claims. "One collector won't heat a home," says Hurrle, despite what some salespeople may tell you.

Of the two types of systems, I personally like the closed-loop systems — collectors with glazed panels — for residential and most commercial applications. If you are heating a warehouse, barn, or less tightly sealed structure that doesn't need much heat, you may want to consider a transpired solar hot air system. They are manufactured by a company called SolarWall.

Transpired collectors for residential structures have been pulled from the market for a number of reasons. As Steve Andrews notes, "Transpired air collectors appear to be very suitable for a range of commercial applications, but seem to present more challenges than opportunities in residential applications — either in existing or new-home applications." One problem with residential applications is that transpired collectors introduce too much fresh air. While fresh air is required for wintertime comfort, these systems can pump thousands of cubic feet of air into a house every hour, resulting in too-frequent air changes. Air pumped into a house forces warm indoor air out through openings in the building envelope (discussed in Chapter 3). The subsequent loss of heat could waste a substantial portion of the heat produced by the collector or your heating system.

Moisture buildup in walls is another potential problem with transpired collectors. Open-loop systems draw outdoor air into a house, bringing a lot of moisture along with it. Because the incoming air has to go somewhere, these systems force the now-moist indoor air through cracks around doors, windows, light switches, electrical outlets, and other openings in the building envelope. The moisture can end up in the insulation in wall cavities, the attic, or the ceiling. As discussed in Chapter 4, moisture that collects in wall insulation greatly reduces its effectiveness. Even a tiny amount of moisture can decrease insulation's R-value (its ability to retard heat flow) by half.

Moisture also promotes mold growth and can, over time, cause wood to rot. Decaying framing members may eventually collapse, resulting in structural damage in the walls of wood-framed houses.

## Conclusion

Solar hot air systems can reduce heating bills and improve home comfort — if properly designed and integrated into a building. Shop carefully. I strongly recommend hiring a professional to install your system. Call solar hot water and solar electric installers in your area to see if they also sell and install solar hot air systems. Research their products and ask to see some installations and talk to their customers.

Before purchasing a solar hot air system, be sure to investigate local building codes and zoning ordinances. You may need a building permit. Also, check out neighborhood or subdivision covenants as well. They may prohibit solar systems (although many homeowners have successfully challenged their homeowner's and neighborhood associations).

If a solar hot air system makes sense for your situation, you will be rewarded many times over. Once you pay off your investment, you'll receive free heat for the life of the system. At that point, you can sit back and enjoy the Sun's free heat and the savings, knowing your work is done. You won't be cutting and hauling firewood to save money on your monthly fuel bill. Nor will you have to worry about the rising fuel costs that are plaguing your neighbors! And, you'll be doing something positive to create a clean, healthy future.

Chapter 9

# Solar Hot Water Heating

Many people are aware of solar hot water systems that heat water for domestic uses — such as showers, baths, and dishwashing (Figure 9-1). These systems are often economical, highly effective, and cheaper over the long haul than conventional water-heating systems. In fact, domestic solar hot water systems represent one of the best buys among *all* the solar technologies. The only solar technology that's more economical is passive solar heating.

DAN CHIRAS

Fig. 9-1: *The author's home in Missouri is equipped with a solar hot water system for domestic hot water, that is, water for showers, baths, dishwashing, and the like. These systems are highly effective and typically designed to supply about 40% to 80% of a family's hot water.*

Solar hot water systems, often referred to as *solar thermal systems,* can also be designed to heat homes. However, solar thermal home heating systems require considerable upsizing and additional changes to generate and store heat to meet demands during extended cold periods.

As the cost of natural gas continues to rise, those who retrofit their homes with solar thermal systems to provide domestic hot water *and* space heat could save a sizeable amount of money and reduce their environmental footprint.

Because solar hot water home heating systems are complex, I'll begin by discussing domestic solar hot water systems (DSHW) designed to heat water for domestic uses. I'll concentrate on those systems that are the most reliable and cost effective. I'll then cover solar hot water systems used for space heating. By the end of the chapter, you should have a good understanding of solar thermal systems and be able to talk with potential installers about the options for your particular situation. For those who want to learn more, I suggest reading Bob Ramlow and Benjamin Nusz's *Solar Water Heating.*

As a friendly reminder, as with all other solar heating options, sealing the leaks in the building envelope, beefing up the insulation, and employing other efficiency measures, such as installing water-efficient showerheads, are the very first steps you should take in your quest for an economical and environmentally sustainable solar hot water heating system.

## Types of Domestic Solar Hot Water Systems

Most domestic solar hot water systems in use today in North America consist of two components: solar panels or solar collectors, and a large solar hot water storage tank (Figure 9-2). The solar collectors are typically mounted on the roof or on the ground in a location that receives full sunlight year round. The solar hot water storage tank is typically located next to the existing water heater so it can feed solar-heated water into its storage tank. In some homes, installers design systems with a single tank. Take a moment to review the system shown in Figure 9-2.

Solar thermal systems such as these are highly effective and, as noted earlier, are typically designed to supply about 40% to 80% of a family's hot water. In extremely warm and sunny climates, a system could satisfy nearly 100% of a family's needs.

As illustrated in Figure 9-2, copper pipes connect the solar collectors with the solar hot water storage tank. This circuit is known as the *solar loop* or the *collector loop*.

As discussed in more detail shortly, a fluid is pumped through the pipes and the collectors where it is heated by the Sun. This fluid can be either water or *propylene glycol*, a type of antifreeze. Propylene glycol never mixes with drinking

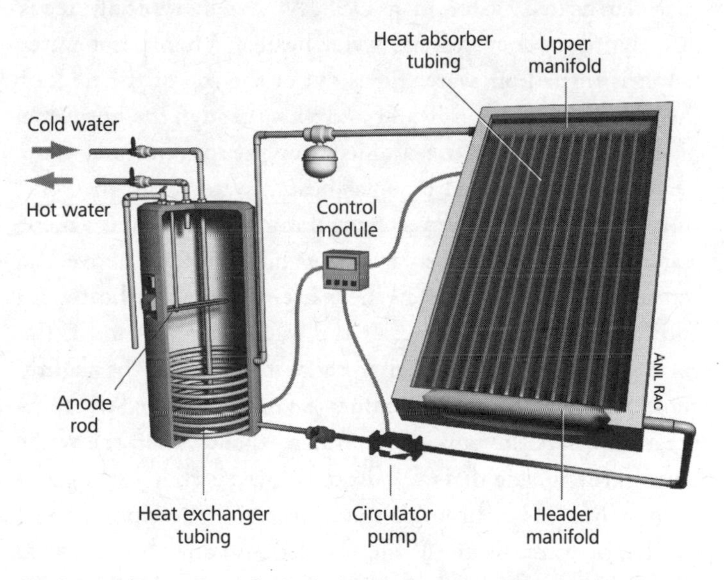


Fig. 9-2: *In this system, a fluid is heated in the solar collectors when the Sun is out. This fluid is then pumped to a water-filled storage tank, located next to the domestic water heater. Heat is stored in the tank (in the form of hot water) until needed.*

water but is nontoxic and therefore safe to use should a leak develop in the system. The glycol solution used in solar hot water systems is designed to be heated to high temperatures without damage, so it can last 20 years or longer if the system is properly maintained. The fluid (water or glycol) that circulates through a solar hot water system is referred to as the *heat-transfer fluid*.

In many solar hot water systems, pumps, sensors, and controls are used to ensure that the system works automatically — starting up in the morning when sunlight warms the collectors and shutting down at night. Systems that require pumps to propel the heat-transfer fluid through the pipes and collectors are referred to as *active systems*. Simpler systems without pumps are known as *passive systems*.

Solar-heated water in a DSHW system typically feeds directly into a conventional water heater. When a hot water faucet is turned on, water flows out of the top of the tank of the conventional water heater and then through the hot water pipes to the faucet. Water drained from the conventional water heater tank is replaced by solar-heated water from the solar storage tank. If the water temperature in the replacement water (that is, from the solar storage tank) is at, or above, the setting on the thermostat in the conventional water heater, no additional heat is required. If the water temperature in the solar storage tank is lower than the water heater's, the slightly cooler water from the solar storage tank will lower the temperature inside the tank of the water heater. When the water temperature inside this tank drops below a certain setting, the burner (in a gas or propane water heater) or heating element (in electric water heaters) will be called into duty, boosting the temperature to the desired setting. Either way, solar hot water reduces the demand for fuel required to heat a family's water.

Solar hot water can also be fed into a tankless water heater. If the temperature of the solar-heated water is at or above the thermostatic setting, the burner in the tankless won't come on. If it isn't, the burner will kick in, raising the temperature of the solar-heated water to the proper temperature.

As noted above, some new solar thermal systems require only one tank. Water in this tank is heated by the heat-transfer fluid that circulates through the solar collectors. The heat it gains in the collectors is stripped off by a heat exchanger. A heat exchanger is a coil of copper pipe located in the wall or in the base of the tank or sometimes outside the tank. It allows solar heat to move from the heat-transfer liquid to the storage tank.

Domestic hot water is drawn directly from the tank. In these systems, the tanks are fitted with one or two backup electric heating elements. If it's cloudy and the solar thermal system cannot generate enough heat to warm the water, the electric heating element(s) kick(s) in, ensuring sufficient hot water.

Single tank systems are easier to install, require less space, and are less expensive. If you have an electric water heater that needs replacement or are installing a system in a new home in which you were anticipating using electricity to heat your water, this may be the system for you. Bear in mind, however, that natural gas is much less expensive and more environmentally sound than electric heat. If your electricity is supplied by a solar electric system, however, the environmental concerns of the electric heating element in the water tank should be greatly diminished. Bear in mind, however, that electric heating elements draw around 4,500 to 5,500 watts and therefore consume a significant amount of energy. Unless your solar hot water system produces most of your domestic hot water, the electrical demand could easily overwhelm a fairly large solar electric system.

### *Direct and Indirect Systems*

Active and passive domestic solar hot water systems can be designed as either direct or indirect systems. Direct systems heat the actual water that occupants of a building consume. That is, the water consumed in a house circulates through the collectors. It is stored in a tank that feeds directly into the hot water pipes of a house. Direct systems are also referred to as *open-loop systems*.

Indirect systems, in contrast, heat a fluid that then heats the water that comes out of the faucet. In these systems, heat-transfer fluid is heated by the Sun in the collector. Its heat is transferred to water in the solar storage tank after it passes through a heat exchanger. As noted earlier, the heat exchanger may be located in the solar storage tank, alongside it, or underneath it. When hot water is required, it flows out of the conventional water heater. Water drained from this tank is replenished by solar-heated water stored in the solar storage tank.

Indirect systems such as this are referred to as *closed-loop* systems. The term "closed loop" refers to the fact that the heat-transfer fluid is in a separate set of pipes and never mixes with drinking water.

## Solar Batch Systems

DSHW systems fit into three broad categories: *solar batch hot water systems, thermosiphon systems,* and *pump circulation systems*. Some of them are active; some are passive. As shown in Table 9-1, solar batch systems are open-loop systems, while thermosiphon and pump circulation systems may be either open- or closed-loop, that is, direct or indirect.

Let's begin with the simplest of these systems, solar batch hot water systems. As shown in Figure 9-3, solar batch hot water systems, or simply batch systems, combine water collection and storage in one unit. That is, the solar collector and storage tank are housed together. Even though solar batch systems are not used in solar space-heating systems, it's a good idea to take a quick look at them. This will help you better understand how solar hot water systems work.

**Table 9-1**
**Solar Hot Water Systems**

| Type of System | Active or Passive | Heat transfer fluid | Propulsive force for heat transfer liquid | Open- or closed-loop | Suitable climate |
|---|---|---|---|---|---|
| Solar batch water heater | Passive | Water | Line pressure | Open | Warm, very infrequent freezing or cold weather |
| Thermosiphon | Passive | Water or propylene glycol | Convection | Open- or closed- with propylene glycol | Warm, infrequent freezing or shut off in winter |
| Pump circulation (Gravity drainback) | Active | Water | AC or DC Pump | Open | Any climate but designed for cold climates |
| Pump circulation (closed-loop antifreeze) | Active | Propylene glycol | AC or DC pump | Closed | Any climate |

Solar batch water heaters are popular in many tropical countries such as Mexico, or desert nations like Israel — areas that remain hot year round. Solar batch water heaters are rather simple devices. In tropical countries, a batch collector usually consists of a single black water tank mounted on the roof of a building. The tanks are painted black, so they absorb more sunlight energy, improving their efficiency. During the day, solar energy heats the water inside the tank. When a hot water faucet is turned on, solar-heated water flows out of the top of the rooftop tank into the hot water supply line that runs directly to the faucets inside the building. Cold water enters the bottom of the tank, replenishing the water drawn out from the tank. Gravity propels hot water down the pipes, and cold water is propelled up into the tank by line pressure. As a result, no pumps are required to move water in this system. Because the water heated in the tank is used inside the building, solar batch hot water systems are classified as direct, or open-loop, system.

Solar batch heaters in use in many more developed countries are considerably more sophisticated than the black roof-mounted tanks you will see in Central America. As shown in Figure 9-3, storage tanks are typically located inside a collector — an insulated box that helps hold in heat at night.

In many countries, the solar batch water heater is usually plumbed into a home's conventional water heater. (That is, it feeds hot water to the conventional water heater.) Therefore, when a hot water faucet is turned on in the house, water is drawn directly from the conventional water heater. Replacement water flows from the solar batch water heater into the tank of the water heater. On hot sunny days, solar batch water heaters produce extremely hot water, so the conventional water heater never kicks on. To prevent scalding, installers place a mixing valve (also called a "tempering" valve) between the hot water line exiting the water heater and the cold water line feeding the collector. If the water leaving the tank is too hot, the valve opens, permitting cold line water to mix with the hot water, bringing the temperature down to a comfortable level.

On cool, cloudy days, when the temperature of the water in the batch heater is below that in the conventional water heater, the conventional water heater turns on, boosting water temperature. The cooler water flowing into the tank from the batch heater is therefore, raised to the desired setting (usually around 115–120° F/46–49° C). On these days, the batch

SOLAR TECHNOLOGY SYSTEMS

Fig. 9-3a: *In this system, water is heated in a tank or series of large pipes inside a collector located on the roof or ground next to a building. Hot water flows directly out of the tank when a hot water faucet is turned on.*

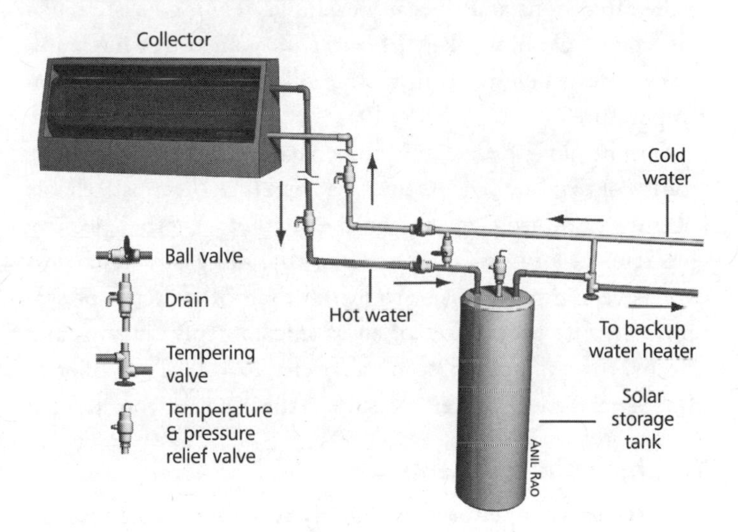


Fig. 9-3b: *This diagram shows the components of a solar batch water heater.*

heater preheats the water that flows into the storage water heater tank. This saves energy because the water from the solar batch water heater is usually much warmer than the water coming in from a well or from the city water supply.

### *Drawbacks of a Solar Batch Water Heater*

Although solar batch water heaters are economical and highly reliable, they do have some drawbacks. One of the most important is that they can only be used in warm climates — areas like Florida that rarely, if ever, experience freezing temperatures. That's because the solar water storage tanks and water lines connecting them to the house are located outside, where they could freeze. Freezing temperatures cause water in the pipes and collectors to expand, creating enough force to crack a copper pipe. When the pipes thaw, water bursts out.

Another problem is that, even in warm climates, solar batch collectors lose heat to the atmosphere at night or on cool, cloudy days. The hottest water is, therefore, typically available in the afternoons and the early evenings. To use a system like this, you would most likely have to shift your use of hot water to the times when water inside the collector is at its maximum temperature.

On the plus side, batch heaters have no moving parts, electronic controls, or sensors and are therefore *the* most reliable solar water heating systems on the market. Another positive consequence of their simplicity is that they are much less expensive and much easier to install than other types of systems. Because they're free of electronic controls, sensors, and pumps, they also require no electricity to run. They operate on line pressure — water pressure in the pipes of your home.

## Thermosiphon Systems

For a system that performs admirably year round, solar designers separate the collection and storage functions. More precisely, they place the collection unit (the solar collectors) on the outside of the house (usually on the roof) where they can absorb solar energy all day long. They place the storage tank inside, where it is much warmer and not subject to freezing. As a result, these systems can operate efficiently even in the coldest weather.

Separate collection and storage systems make up the bulk of the solar hot water systems in use today. The most basic of these is known as a *thermosiphon system* (Figure 9-4). It consists of a collector, a storage tank, and pipes connecting the two. Instead of relying on a pump to propel solar-heated water from the collector to the storage tank, water flows by convection. (Convection occurs when a fluid is heated. The heated fluid expands, becomes lighter, and thus naturally rises.)

Figure 9-4 shows the anatomy of a thermosiphon solar hot water system. As illustrated, the heat-transfer liquid in the collector is warmed by sunlight. It then rises and flows out of the

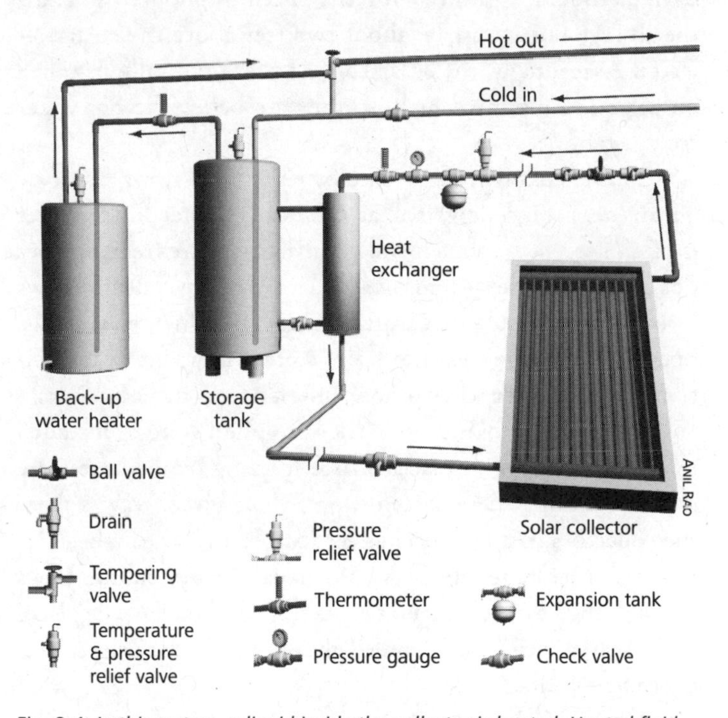


Fig. 9-4: *In this system, a liquid inside the collector is heated. Heated fluid expands, becoming lighter, then rises, creating a thermosiphon. That is, it draws cooler liquid in from below, creating a convection loop.*

solar collector to the solar storage tank. The flow of heated liquid out of the solar collector, in turn, draws cooler fluid into the bottom of the collector.

In these systems, heat "pumps" the fluid from the collector to the storage tank. This natural pumping action is created by convection, and is known as a *thermosiphon pump.* Convection continues to pump fluid through the system so long as the fluid in the solar collector is hotter than the fluid in the bottom of the solar storage tank.

The convective flow of liquid in this system is a simple, non-mechanical thermosiphon pump that operates during daylight hours. However, for the thermosiphon to operate, the storage tank must be about two feet above the solar collector. Accordingly, the solar collectors are typically installed on a ground-mounted rack, so they are below the hot water storage tank.

Thermosiphon systems employ either water or a nontoxic antifreeze, propylene glycol, as the heat-transfer fluid. Water is often used in systems installed in areas where freezing does not occur. In these systems, water heated in the collector flows directly into the solar hot water storage tank. In open systems, hot water from the solar hot water storage flows into a conventional water heater when a hot water faucet is turned on.

Propylene glycol is used in thermosiphon systems installed in colder climates — places where freezing temperatures are encountered. In these systems, propylene glycol travels from the collectors to a heat exchanger located in or near the solar hot water storage tank. As the solar-heated liquid flows through the heat exchanger, heat is transferred from the glycol to the water in the tank. Cooled glycol flows back to the collector to be reheated.

Open-loop thermosiphon systems (those that use water as the heat-transfer fluid) are relatively simple, inexpensive, and trouble-free. Closed-loop systems (those that use propylene glycol as the heat-transfer fluid) are more complicated, in large part because of the addition of a heat exchanger.

## Pump Circulation Systems

The most widely used solar hot water systems in developed countries are pump circulation systems. They are similar to the thermosiphon system, except that the driving force — the force that moves the heat-transfer liquid — is a small electric pump.

Solar collectors in pump systems are usually mounted on roofs, and less commonly on the ground near the building. The solar hot water storage tanks are located inside the house, next to a conventional water heater — typically a storage water heater. An electric pump propels the heat-transfer liquid up through the collectors where it is heated by the Sun. The heated fluid then flows back to the thermal storage tank, where the heat is deposited.

As in thermosiphon systems, pump-driven systems use either water or propylene glycol as the heat-transfer fluid. If water is used, the system can be designed as an open-loop system — that is, a direct system. Although open-loop systems have their advantages, freezing may occur in cold climates. To prevent pipes from freezing and bursting, designers developed a gravity drainback system, simply referred to as a *drainback system*. Figure 9-5 shows a drainback system.

## Gravity Drainback Systems

To understand how these systems work, let's begin at sunrise. As the Sun heats the solar collectors, a temperature sensor located at the top of the collector (see Figure 9-5) activates a pump near the storage water tank. It pumps water out of the storage tank through the solar collectors. As water flows through the collector, it warms up. As water cycles through the solar loop between the storage tank and the collector, it gets hotter and hotter.

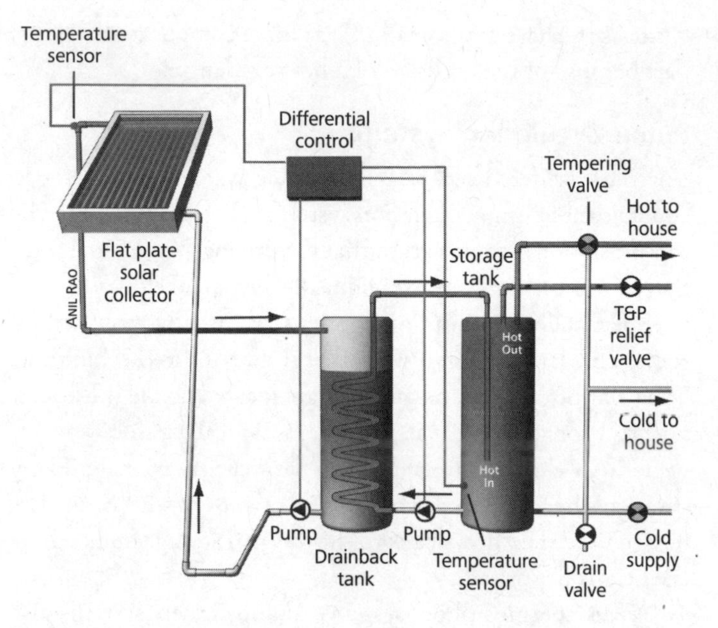


Fig. 9-5: *In this drainback system, water is the heat-transfer fluid. It circulates through the system thanks to a small electric pump.*

Another pump, also shown in Figure 9-5, circulates water out of the conventional water heater (right) through a heat exchanger located in the solar hot water storage tank. Heat stored in the latter is thus transferred from solar-heated water to household water (in the conventional water heater).

Water continues to circulate through both loops of the system (the collector loop and storage tank loop) until the Sun goes down. At that point, the system shuts down. The pumps turn off. To prevent water from freezing and damaging pipes, however, all the water in the pipes and collectors drains out of them. It empties into the solar storage tank by gravity — giving the system its name, *gravity drainback*.

The pumps in drainback systems are controlled by a controller, a small logic circuit that is wired into a temperature sensor located in the solar collectors. When the Sun goes down, the temperature of the liquid in the collectors immediately begins to fall. The sensor then sends a signal to the controller, and the controller switches off the circulating pumps.

### *Pros and Cons of Gravity Drainback Systems*

One advantage of gravity drainback systems is that they can be used in all climates — even extremely cold ones. They're a bit simpler and therefore less expensive than glycol-based active systems. Another advantage is that they do not require propylene glycol. As you shall soon see, propylene glycol may deteriorate due to high temperatures if the system sits idle for long periods (for example, while a family is on a prolonged vacation or if the pump breaks down and is not repaired quickly). When this occurs, the glycol needs to be drained and replaced. Glycol currently costs about $50–$60 per gallon, and systems require three to five gallons, depending on the length of the pipe run and the number of collectors. Glycol is typically replaced by a trained professional, which adds to the cost of this maintenance.

On the downside, drainback systems require the largest pumps of all solar hot water systems. In addition, although they are designed to drain when the pumps stop, there is still risk of damage from freezing. So, they may not be the best option for the coldest of climates. In such instances, installers may use a mixture of propylene glycol and water to provide a greater degree of freeze protection.

Another downside is that gravity drainback systems aren't good in areas with hard water. Calcium deposits in the pipes can clog the system. Distilled water should be used in the solar loop, but it requires special fittings.

Another disadvantage of drainback systems is they can't be used on buildings taller than two stories because the pumps can't move water any higher.

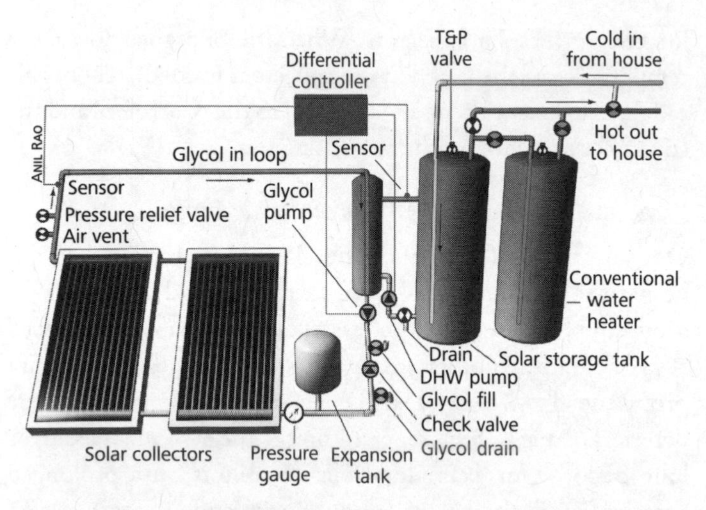


Fig. 9-6: *In this pump-driven, glycol-based system, propylene glycol is the heat-transfer fluid. It circulates through the collectors. Heat is transferred from this fluid to the water in the storage tank via a heat exchanger.*

## Closed-Loop Antifreeze Systems

Although drainback systems work well in cold climates, most active systems are glycol-based pump-driven systems. These systems require a heat exchanger to transfer heat from the glycol to the water inside the solar hot water storage tank. As noted above, heat exchangers may be located inside the solar hot water storage tank, alongside it, as shown in Figure 9-6, or underneath, it. These systems also require components not found in other DSHW systems, for example, an expansion tank as shown in Figure 9-6. Glycol expands when heated. The expansion tank accommodates the hot glycol, preventing pressure from building to dangerous levels in the solar loop.

### *Pros and Cons of Glycol-Based Systems*

Closed-loop antifreeze systems are popular and reliable. They work well in all climates, and in cold climates they provide excellent protection against freezing. On the downside, closed-loop systems are the most complex and therefore the most costly of all solar hot water systems. In addition, they have more parts and, therefore, there are more chances for things to go wrong. As noted earlier, the propylene glycol may need replacement if the glycol routinely overheats, which may occur if the circulating pump breaks down or a family is away from the home during the summer for a protracted period. During such periods, hot water is not removed from the solar storage tank. This causes the system to shut down, so the glycol doesn't circulate through the collector loop. Glycol that sits in the pipes in the collector overheats and turns black and viscous. One way to prevent this problem is to program the controller to circulate in reverse at night. This bleeds heat out of the storage tank so the system continues to operate (glycol circulates through the collectors during the day).

## Solar Collectors

Now that you understand how solar hot water systems operate, let's take a brief look at the solar collectors. There are two types of collectors used in solar hot water systems: *flat-plate* and *evacuated-tube*.

### *Flat-Plate Collectors*

The flat-plate collector consists of a glass-covered insulated box. Inside are copper pipes attached to a flat absorber plate — all coated by a highly absorbent black material, known as a *selective surface*. A selective surface is so named because its color and surface texture allow it to absorb sunlight much more efficiently than conventional black surfaces. Because of this, collectors with selective surface absorber plates are more efficient — they can convert more solar energy into heat.

Sunlight consisting of visible light and heat (infrared radiation) enters a flat-plate collector (Figure 9-7). Visible light is absorbed by the dark-colored absorber plate, where it

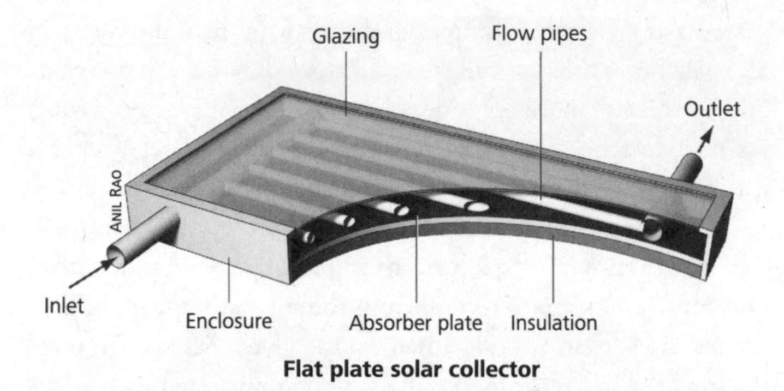


Fig. 9-7: *Propylene glycol or water circulates through the pipe in this panel, removing heat from the collector and transferring it to a water storage tank.*

is converted to heat. Heat waves (infrared radiation) heat the collectors directly.

Flat-plate collectors are the most commonly used of all collectors, and they are useful for applications that require water temperatures under 140° F (60° C), notably domestic hot water systems. They're also useful for many space-heating applications, notably forced hot air and radiant floor heating systems. Unfortunately, they don't achieve the high temperatures required for baseboard hot water systems.

## *Evacuated-Tube Collectors*

An evacuated-tube collector consists of 20 to 30 long, parallel glass or plastic tubes. Inside each tube is a copper pipe connected to a copper fin to increase its surface area. Both are coated with a selective-surface material to increase solar absorption and heat production. When sunlight strikes the collector, heat is absorbed by the pipe and fin and is transferred to the heat-transfer fluid inside the pipe.

Evacuated-tube collectors get their name from the fact that air is pumped out of the glass or plastic tube, creating a vacuum.

Interestingly, vacuums are great insulators. (That's why thermos bottles work so well.) The vacuum helps hold solar heat inside the collector, even on cold days, which improves their efficiency and permits greater solar heat gain.

Evacuated tubes operate, in part, passively. When sunlight strikes the collector, a fluid inside each copper pipe flows upward by convection (passively) to a heat exchanger located at the top of the collector (Figures 9-8 and 9-9). Here, heat is transferred to another heat-transfer fluid that is actively

DAN CHIRAS

Fig. 9-8: *Glass tubes contain a heat-absorbing pipe and fin that absorb the Sun's energy, transferring it to a heat-absorbing liquid. When heated, it rises to the top of the unit, to a heat exchanger, as shown here.*

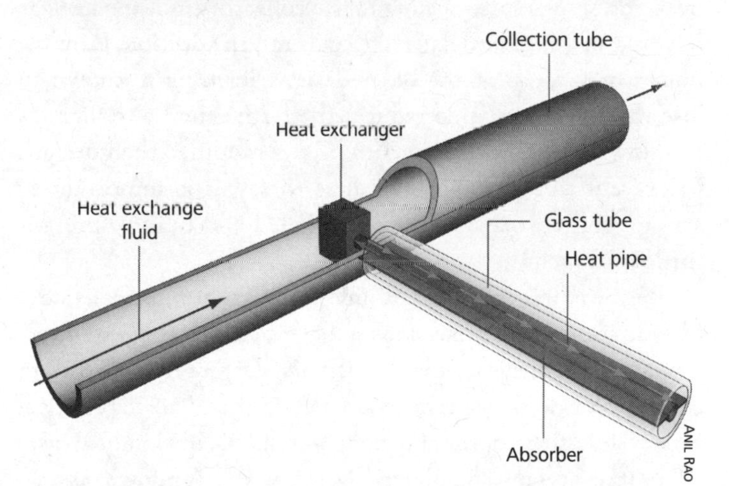


ANIL RAO

Fig. 9-9: *Heat collected by each tube is transferred to propylene glycol flowing through the heat exchanger, as shown here.*

pumped through the system. It carries the heat to a solar hot water storage tank, where it is stored for later use. The cooled fluid returns to continue the cycle, continually drawing heat off the collector tubes.

Evacuated-tube solar collectors perform well in many climates. Manufacturers routinely claim that they outperform standard flat-plate collectors by absorbing and retaining more sunlight energy. This, in turn, increases their efficiency and hot water production. Bob Ramlow, a solar thermal water system expert and senior author of *Solar Water Heating*, pointed out to me that "there's a lot of hype about evacuated-tube solar collectors," making it imperative to consider their pros and cons carefully. For example, while the vacuum in a collector acts as an insulator that helps reduce heat loss from the *inside* of the collector tubes and therefore increases hot water production, in regions with frequent snows, evacuated tube solar collectors shed snow poorly. This is especially true when the collectors are mounted parallel and close to a roof. Because they retain heat, the snow tends to cling to the collectors, not melt off, as it would in a standard flat-plate collector. In addition, Ramlow points out, some of the older designs have been known to lose their vacuum. This reduces their efficiency, although, as Ramlow says, some manufacturers have modified their designs to prevent this problem. Needless to say, it is important to select a collector that has been designed and built to prevent the loss of vacuum.

Besides these problems, Ramlow notes that numerous side-by-side comparisons have shown that evacuated-tube collectors do *not* outperform flat-plate collectors, despite manufacturers' claims. Nor do they perform any better than flat-plate collectors in "less-than-optimal regions" — that is, in cloudy areas.

Not everyone completely agrees with Ramlow's assessment. Evacuated-tube solar collectors do perform better than flat-plate collectors in many circumstances, says Henry Rentz, owner of Missouri Valley Renewable Energy. He has installed solar hot water systems for a dozen years. David Sawchak of Morningstar Enterprises, who has been in the solar hot water business since the 1970s, agrees with Rentz. They're especially useful for high-temperature applications, like baseboard hot water systems. As noted above, baseboard hot water systems require higher temperatures than either forced-air space-heating systems or radiant floor heating systems.

Both experts point out, however, that it is important to consider your choices very carefully. Be sure to consult with a local solar hot water expert — one who really knows what he or she is talking about — to determine if an evacuated-tube solar hot water collector makes sense in your area and for your application.

Ramlow notes that evacuated-tube collectors will produce water as hot as 220° F, compared to 180° F for flat-plate collectors. Although this may seem advantageous, water can boil in the storage tank (typically, when families go on vacation during the summer). Rarely will water boil in a flat-plate collector system. To protect against this, you should consider installing a larger storage tank if you install evacuated-tube collectors.

"When investing in evacuated-tube collectors, a buyer must pay particular attention to quality because while some of the highest rated collectors are evacuated-tube type, most of the lowest rated collectors are also evacuated-tubes," Ramlow says.

## Solar Hot Water Space-Heating Systems

Now that you understand which domestic solar hot water systems are typically installed and how these systems work, it's time to take a look at solar hot water heating systems. Solar hot water heating systems are larger and a bit more complex than domestic solar hot water systems. While a domestic solar hot water system — designed to provide hot water for showers, dishwashing, and the like — typically has two to four

DAN CHIRAS

Fig. 9-10: *This space-heating system on an architectural firm in Colorado Springs uses numerous evacuated tube solar collectors. One of the main problems with systems like these is that they produce a surplus of hot water during the summer.*

collectors, a system designed for space heating typically uses eight or more solar collectors Figure 9-10. These systems also employ a much larger storage tank; instead of 90 gallons, they may require 180 up to 1,000 gallon storage tanks. Large tanks are needed to store the large quantities of solar-heated water required to heat buildings.

Solar hot water space-heating systems are active systems. That is, they are pump-circulation systems. These systems may be either drainback or glycol-based. As just noted, solar heat is stored in water tanks; though it can also be stored in beds of sand. I'll start with water storage systems, as they are the most common type.

### *Drainback Solar Hot Water Heating Systems*

In a drainback solar hot water space-heating system, the design includes one large storage tank or a single drainback tank plus one or more storage tanks to hold hot water for times of need. Most systems have one large, extremely well-insulated tank (typically 200 to 1,000 gallons). This tank is typically filled with distilled water. As Bob Ramlow notes, be sure to use distilled water in the collector loop. Never use tap water. It contains minerals that will deposit on the inside walls of the pipes and collectors, obstructing flow and reducing the efficiency of the system.

On sunny winter days, the water in the storage tank circulates through the collectors, becoming hotter and hotter. However, when clouds obscure the Sun or the Sun sets, the system shuts down. All the water in the collectors and pipes drains back into the storage tank. If the home requires space heat, hot water inside the solar storage tank is there to provide it.

To understand how heat is stripped out of hot water inside the storage tank, consider an example. Imagine we've installed a solar hot water space-heating system in a house previously heated by a forced-air furnace. In such systems, a furnace heats room air passing through it, using natural gas, propane, oil, or electricity. The hot air is circulated through supply ducts in the house. Cold room air returns to the furnace via cold air returns (ducts that run through walls, attic spaces, and under floors). When the cold air reaches the furnace, it is reheated.

When a solar hot water system is added to a home such as this, when space heat is required, heat is drawn principally from the solar storage tank if the water inside the tank is sufficiently hot. The furnace serves as a backup. Here's what happens: when the thermostat in the house calls for heat, solar-heated water flows out of the storage tank through a heat exchanger installed inside the duct that normally carries hot air out of a furnace. The furnace fan turns on and blows cool return air over the heat exchanger. The cool air blown over the surface of the heat exchanger strips the solar heat from it. The solar-heated air is then distributed through the house via the existing supply ducts in the house. The solar storage tank continues to provide heat, circulating hot water through the heat exchanger in the furnace, so long as it is needed, and so long as there's enough hot water in the tank. If the water temperature inside the tank drops below the desired setting, the gas furnace turns on, taking over. As noted earlier, it provides backup heat to the solar system. All this is controlled by a logic circuit, a small computer. It is fed temperature information from the thermostat and sensors located at strategic locations in the system, for example inside the solar storage water tank.

Drainback systems work well in solar home heating systems, supplying space heat *and* heating water for domestic consumption. Bear in mind, however, an additional heat exchanger is needed to provide heat to a family's conventional water heater (or tankless water heater).

## Glycol-Based Systems

Glycol-based systems tend to be the system of choice for space heating, especially in cold climates. In these systems, solar heat can be stored in a single large tank, in multiple storage tanks, or in beds of sand from which the heat is later extracted and delivered to the home's heating distribution system — either ducts or pipes.

Designs of closed-loop glycol-based space-heating systems are identical to those described for domestic solar hot water. However, as in drainback systems, most solar home heating systems employ a single large, extremely well-insulated water tank located in a basement to store solar-heated water. This tank is heated by the solar-heated glycol that circulates through the collectors. The heated glycol circulates through the collectors and then through a heat exchanger inside the storage tank, where the heat the glycol has absorbed is transferred to the water.

Heat drawn from the storage tank is distributed via pipes to a heat exchanger in forced-air systems, as described above.

It can also be circulated through pipes in the floor in radiant floor systems, or pipes that lead to baseboard heaters in baseboard (hydronic) hot water systems. Heat is drawn out of the solar storage tank via a heat exchanger.

If the system is designed for domestic hot water, too, a separate heat exchanger is installed inside the storage tank. It feeds hot water to a domestic water heater — either a tankless water heater or a conventional storage water heater.

Pumps in glycol-based systems can be run by household current or by electricity produced by a PV module mounted near the hot water collectors (Figure 9-11). This is known as *direct PV,* because the photovoltaic module is wired directly to the pump that circulates glycol through the solar loop. Although PV modules work well, they have some limitations. One of the most significant is that they are unable to lift the heat-transfer fluid more than a single story. You won't be able to install one on a two-story home. When in doubt, ask a local expert for his or her opinion.

One of the main disadvantages of solar hot water space-heating systems is that they are sized to produce heat in the winter when solar gain is minimal and heat demand is high. Throughout the rest of the year, however, the demand for heat is minimal — or nonexistent. Consequently, solar thermal systems designed for space heat produce a lot more hot water than is required. To prevent water in the storage tanks from boiling in the off-season — late spring, summer, and early fall — designers typically install a dump load — a place to dump excess heat during the summer. Excess heat can be diverted to

DAN CHIRAS

Fig. 9-11: *This PV module powers only the circulation pumps in the solar hot water system, not the fans or pumps required to distribute heat through a home's heating system.*

pipes buried around the perimeter of a home. Heat flows out of the pipes into the ground, warming it. If designed correctly, this heat can move into a home — through the basement walls — during the winter, keeping it warmer. Insulation buried over the pipe helps retain heat, making it available during the winter.

Heat can also be dumped into hot tubs or swimming pools to lengthen the swim season. In addition, heat can be dumped in very large and extremely well-insulated stainless steel storage tanks buried in the ground. Some of this heat can be drawn off in the winter by heat exchangers and used to heat the home in the winter. Alternatively, heat can be dumped into insulated sand beds in the ground around or under a home. To learn more about heat dumping, check out Chuck Marken's article "Overcoming Overheating," published in *Home Power*, 142, 2011.

With this downside in mind, my advice is to think long and hard about installing a solar hot water space-heating system. As you have seen, they are useful only during the heating season. If you are in an area that has a long heating season, the system will make a lot more sense than if you live in an area with a relatively short heating season.

## A Word on Storage Tanks

Storage tanks for solar hot water space-heating systems should be well built and durable. They must be made of materials capable of storing water for long periods at 180°F or higher (if evacuated tube collectors are installed). High-temperature fiberglass tanks work well as do insulated stainless steel dairy tanks. Beware of homemade steel storage tanks and wood-framed tanks lined with plastic or rubber roofing materials (EPDM). They may be inexpensive to build, but they don't last very long. Ordinary storage water heaters aren't such a good idea either. Conventional water heater tanks are damaged by high-temperature water.

Ideally, tanks should be seamless and jointless to reduce the possibility of leakage. They also have to be sized so they can be carried into a home without cutting out a wall! Larger fiberglass tanks come in pieces so they can be assembled inside the building.

For years of leak-free performance, it is recommended that pipes enter and exit at the top of the tank, rather than from fittings located along the bottom. If you want to learn more about tanks, I recommend Bob Ramlow and Ben Nusz's book, *Solar Water Heating*. It has an excellent chapter on solar space heat. (Tom Lane's book, *Solar Hot Water Systems,* is another excellent, although much higher level, book — great for those who are thinking about getting into the business.) You can also consult with installers in your area.

As noted earlier, it is also possible to store solar heat in sand beds. To do this, hot water is circulated through pipes embedded in insulated sand beds, often under the concrete slab of a home. These systems do work, if designed and installed correctly. Ramlow and Nusz's book discusses them in detail.

Another option for heat storage, though, is to pump heat into the concrete slab of a building. The heat is delivered directly from the collectors into a set of pipes embedded in the slab during construction. These systems provide radiant floor heat and work well in extremely cold climates. Be sure to insulate under the slab and around the perimeter of the foundation. During the summer, you may need to divert heat to a dump load, for example, pipes buried in the ground outside to prevent overheating, or a large insulated storage tank buried in the ground under or alongside the house.

## Choosing the System to Meet Your Needs

Despite the many options, it's not difficult to select a solar hot water heating system that will work for you. If you live in a freeze-free climate, you should consider a drainback system —

an open-loop, pump circulation system. Remember that open-loop drainback systems are not generally recommended in regions with hard water because minerals in hard water can deposit on the inside of pipes in the solar loop. Over time, mineral deposits can reduce flow rates and the efficiency of a system. In colder climates or in regions with hard water, your best bet is a closed-loop glycol-based system.

## Conclusion: Sizing and Pricing a Solar Hot Water Heating System

Before sizing a solar thermal home heating system, as with any system, it is important to make your home or business as efficient as possible. How large a system you need depends on many factors, including wintertime temperatures, the amount of sunshine you receive, how energy-efficient (airtight and well insulated) your home is, and temperature requirements (do you require the temperature cranked up to 80° F to stay warm?).

Your best bet is to call local installers who will assess your heating requirements, the energy-efficiency of your home or business, solar availability, and other factors, then recommend a system size. A small home-sized system for an efficient home in a sunny climate could cost as little as $12,000; a larger system for an energy-inefficient home with poorer solar resources could cost many times more — $50,000, even more.

The economics of the investment, however, are generally quite favorable — that is, these systems are usually a good investment. The return on investment depends in part on the type of heating system you are currently using. If you are replacing electric heat with solar heat, the investment is almost always sound — as it is extremely costly to heat with electricity. Replacing propane heat with solar heat can also be quite cost effective, too. Replacing natural gas with solar heat may have a lower return on investment, although natural gas costs have increased dramatically in the past ten years and are expected to continue to rise.

Don't forget that in the United States there are economic incentives from the federal government for solar hot water systems. You can receive a 30% federal tax credit for installing a solar hot water system.

State or local incentives may also be available. A local installer can give you the rundown on state, local, and federal incentives. Or, you can check www.dsireusa.org to see what's available in your state.

Chapter 10

# Energy-Efficient Backup Heat

The goal of this book is to familiarize readers who want to increase their reliance on solar energy for heating with their options, to even help readers find ways to utilize the Sun's energy to produce nearly 100% of the heat they need for their homes or businesses. Although your home or business can be heated entirely by clean, renewable solar energy, it's easier to achieve this goal when starting from scratch. When retrofitting a building, your options are limited, and heating the building entirely with solar energy is much more difficult. In such cases, you'll likely need a backup heating system.

If you are retrofitting an existing building with solar, your existing heating system could become your backup heat source — playing second fiddle to the energy-efficiency measures you incorporate and the passive solar, solar hot air, or solar hot water system you install. If your heater is old and inefficient and therefore in need of replacement, you may want to consider installing a new, more energy-efficient unit.

If you are building a new energy-efficient solar home from scratch, you'll want to install an energy-efficient backup system — if for no other reason than to comply with the local building code. Building codes require backup heating to prevent pipes from freezing.

This chapter explores energy-efficient backup heating system options for new and existing homes, including energy-efficient furnaces and boilers.

## Home Heating Basics

The vast majority of homes in North America are heated by the combustion of fossil fuels — natural gas, propane, or fuel oil — or by electricity, which, of course, is generated by coal-fired power plants, natural gas plants, or nuclear power plants. The vast majority of homes in North America are equipped with forced-air heating systems. They're the least expensive of all mechanical heating systems to install, hence their popularity. Unfortunately, they tend to be the least efficient of all your choices.

### *Conventional (Low-Efficiency) Furnaces*

In forced-air heating systems, hot air produced by the furnace is distributed through the home by a large fan, known as a *blower*. The blower propels the heated air through an extensive duct system that supplies heated air to each room (Figure 10-1). Cool room air then returns to the furnace for reheating by a separate set of ducts, known as the *cold air return ducts*. (Ducts in forced-air systems are notoriously leaky, so it is a good idea to have yours tested and seal the leaks with mastic, a white gooey paste that provides the best, and longest-lasting seal.)

Furnaces are usually installed in a basement, or less commonly in a utility room or a well-vented closet. Gas furnaces contain a combustion chamber where natural gas or propane is burned. The burner may be ignited by a pilot light, a small flame that burns 24 hours a day, or by an electronic ignition, which provides a spark to ignite the fuel when heat is required. (The latter is more efficient.)

In a conventional furnace, referred to as a *natural-draft* furnace, heat produced inside the combustion chamber is

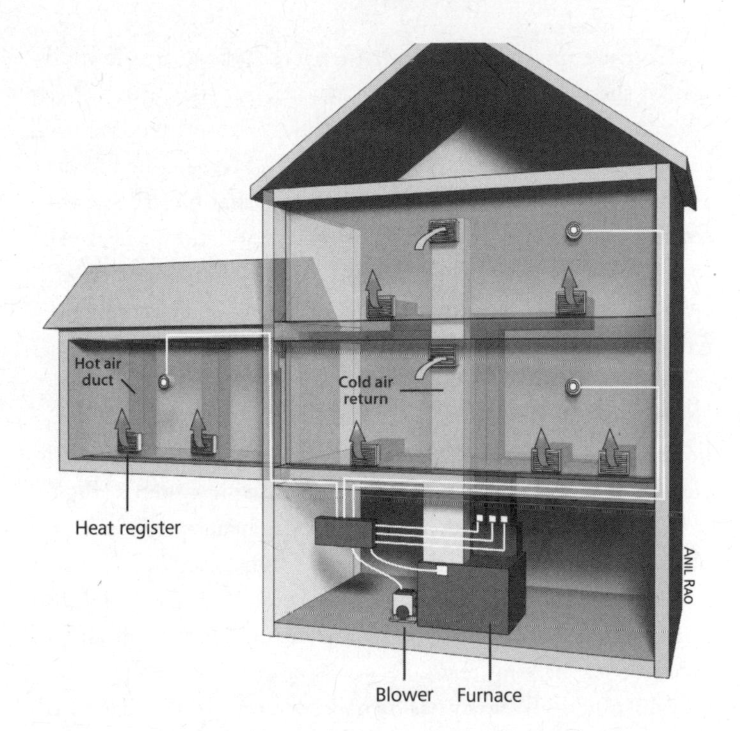


Fig. 10-1: *Heat generated by the furnace is distributed through a duct system.*

transferred to cool air flowing over its heat exchanger. The blower propels the newly heated air through the supply ducts to the rooms. Cool air then flows back to the furnace through the return duct system.

Combustion gases that contain potentially lethal pollutants are vented from the combustion chamber to the outdoors through a flue pipe. As the hot gases rise, they create a partial vacuum in the combustion chamber. This draws room air into the fire, ensuring a continuous supply of the oxygen needed for proper combustion. The rise of hot air through the flue, together with the inflow of room air, is known as *draft*. (Draft is created by convection.)

Conventional natural-draft furnaces have been the mainstay of the heating industry for decades, but they are the least efficient of all furnaces on the market. Those manufactured before 1992 have efficiencies below 78%. Many old furnaces are only 55% to 65% efficient. Translated, that means that they convert only 55% to 65% of the fuel they burn into heat. The rest is wasted. If you were buying bananas at the same efficiency, you'd buy ten, then throw five and a half to six and a half away as soon as you got home.

### *High-Efficiency Furnaces*

Fortunately, numerous manufacturers now produce high-efficiency furnaces for use with forced-air heating systems (Figure 10-2). The higher the efficiency, the more heat you get per unit of fuel that's burned in the furnace. High-efficiency furnaces not only produce and deliver a lot more heat from the fuel they burn than older models, they save an enormous amount of money over the long-term.

Most high-efficiency gas furnaces are referred to as *induced-draft models*. That means that they contain a low-wattage electric fan that draws air from *outside* the home into the combustion chamber — rather than drawing air "naturally" from inside the home, as in older, natural-draft furnaces. The fan also propels exhaust gases from the combustion chamber out of the house via the flue pipe.

Induced-draft furnaces are equipped with more efficient heat exchangers and electronic ignitions, which eliminate the need for a standing pilot light. Both devices increase operating efficiency. Many high-efficiency furnaces contain controls that allow the blower to continue to blow hot air over the heat exchanger after the thermostat shuts the furnace down. This enables the blower to strip more heat from the heat exchanger and internal components of the furnace, increasing the efficiency of the furnace even more.

COLEMAN

Fig. 10-2: *High-Efficiency Furnace.*

The most efficient gas furnaces on the market today are known as *condensing furnaces*. These furnaces contain a second heat exchanger that extracts heat from the exhaust gases — heat that would normally be lost as it escapes through the flue. By extracting heat from these gases, the heat exchanger cools the gases until the moisture (water) contained in the gases condenses. The condensation of water vapor from air also — almost mysteriously — releases heat. Condensing furnaces capture this heat as well and distribute it throughout the building as needed. Because so much heat is removed by the heat exchangers in condensing furnaces, waste gases can be vented through plastic pipe. When the furnace is operating, the flue pipe is barely warm to the touch. The moisture removed from the exhaust gases, however, must be drained into a nearby floor drain.

Both condensing and non-condensing furnaces are equipped with sealed combustion chambers. This feature prevents dangerous exhaust gases, such as carbon monoxide and nitrogen dioxide, from entering room air. Installing an energy-efficient furnace or replacing an old gas furnace with a superefficient model, therefore, will not only save you money, it could improve indoor air quality in your home.

When shopping for a new furnace, contact local installers for advice. If they're reputable, they'll tell you which models are the most efficient and reliable. You can also look on the Energy Star website for Energy Star-qualified models. An Energy Star label on a furnace ensures that it is among the most efficient on the market. Still, shop carefully. Among this elite group of furnaces you can still find considerable variation in energy efficiency. In other words, some Energy Star furnaces are a lot more efficient than others.

Furnace efficiencies are listed as annual fuel utilization efficiency, or AFUE, for short. They range from 83% to 99%. As a rule, the induced-draft furnaces have efficiencies in the 80% range, and induced-draft condensing furnaces are higher — in the 90% range — with some as high as 99%. Be wary of brand new models. It's best to install a furnace that's been on the market for at least four or five years — just to be sure that the bugs, if any, have been worked out. Check out the warranty very carefully. Deal with a company that's been in the business for at least five years, preferably more.

To view a list of Energy Star-qualified furnaces, visit the Energy Star website (www.energystar.gov), and click on "Products," then "Heating and Cooling," and "Furnaces." Be sure to check out the product lists.

If you can't afford a new furnace, you should consider modifying your existing furnace to improve its efficiency. To learn how, read the Department of Energy's Consumer's Guide at their Energy Efficiency and Renewable Energy website at: www.eere.energy.gov/.

## *High-Efficiency Oil Furnaces*

Fuel oil is one of many products extracted from crude oil or petroleum. Fuel oil is not anywhere near as thick as crude oil. It's more like diesel fuel.

Fuel oil is burned in furnaces in many parts of the United States — notably the Northeast, the Northwest, and the central states. Fuel oil is injected into the furnace's combustion chamber through a nozzle that produces a fine spray of tiny droplets.

If you have an oil furnace manufactured prior to 1992, chances are it's quite inefficient — possibly only 50% to 60% efficient. In contrast, newer Energy Star-qualified oil furnaces have efficiencies in the 83% to 86% range. Even more efficient are condensing oil furnaces, which are about 95% efficient. Although condensing models are more efficient, they're not very common and are often not recommended, mainly because fuel oil contains many chemical contaminants (such as sulfur) — far more than are found in natural gas or propane. These contaminants condense out and produce a fairly corrosive liquid that can damage the internal components of a furnace.

Although sealed combustion chambers are common in high-efficiency gas furnaces, very few high-efficiency oil furnaces come with this feature. The reason is that cold outside air drawn into the combustion chamber in an oil furnace reduces combustion efficiency. It can also impair ignition.

Because oil furnaces are not usually equipped with sealed combustion chambers, carbon monoxide can leak into room air. Carbon monoxide is a colorless, odorless, poisonous gas that can kill people and animals when present in high concentrations. It is therefore a good idea to install a carbon monoxide detector or two if you heat with an oil furnace. One detector is recommended for every floor, in a location where they can be heard, especially near the master bedroom. Check building code requirements to be certain you have the right number and they're in the correct location.

## Conventional Boilers and Hydronic Heating Systems

If your home is heated by a radiant floor or baseboard hot water system, heat is generated by a boiler (so named because it boils water). The heated water is then circulated through pipes, supplying heat to in-floor, baseboard hot water, or other types of radiators (heat exchangers). These systems are referred to as *hydronic heating systems*.

Boilers are typically located in basements or utility rooms, just like furnaces. Also like furnaces, they typically burn natural gas, propane, or heating oil. The combustion of fossil fuels in a boiler heats water that is circulated in pipes surrounding the combustion chamber. In some cases, electricity is used to generate heat, although electric heat can be quite expensive.

In homes with radiant floor systems, also known as *in-floor heating*, hot water generated by the boiler is pumped through pipes beneath the finished floor — either in a concrete slab or under wood flooring or tile (Figure 10-3). Heat is released and absorbed by the floor, which warms up, radiating heat into the rooms.

In homes equipped with baseboard hot water systems, water is circulated throughout the house to baseboard radiators. These devices are heat exchangers that release heat into the rooms in which they are located (Figure 10-4). Installed along the base of walls, they contain a series of aluminum fins attached to a copper pipe that runs the length of the unit. Heat flowing through the copper pipe flows into the fins, which dramatically increase the surface area for heat exchange. The heat then radiates into the rooms. The fins also heat the air above them. Air heated by the fins expands, becomes lighter, then rises, moving via convection through the rooms of a house.

In-floor and baseboard heating systems are inherently more efficient than forced hot air systems and therefore more cost effective. Their efficiency stems from the fact that forced-air heating systems tend to pressurize the interior of buildings, that is, increase pressure due to the flow of hot air into rooms via the duct system. Pressuring the interior forces air out through tiny leaks in the building envelope.

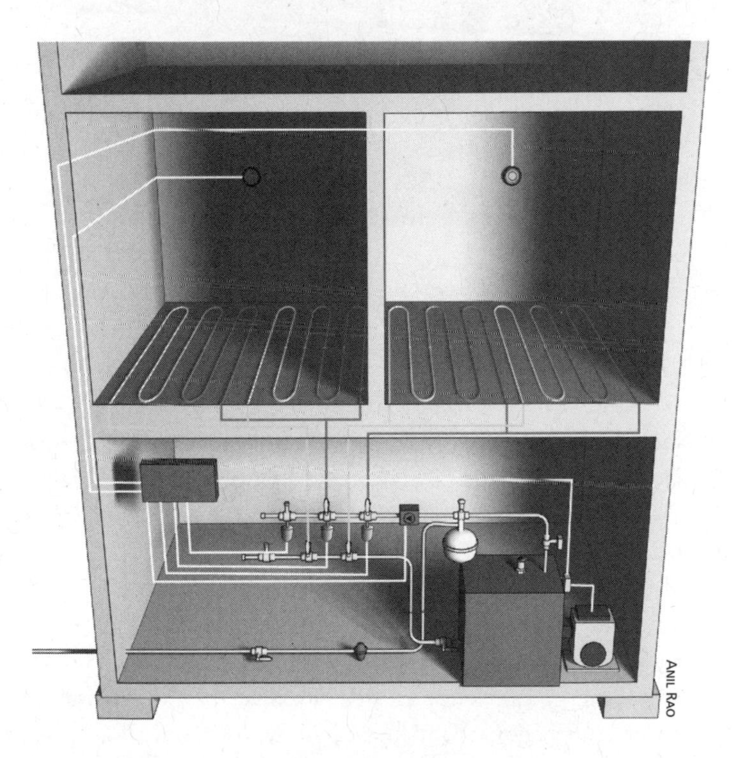


Fig. 10-3: *Radiant Floor Heating System.*

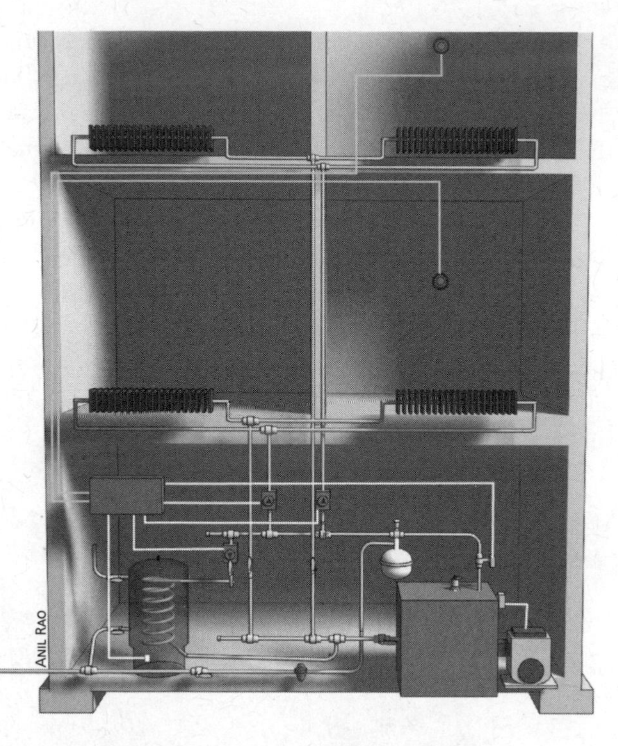


Fig. 10-4: *Baseboard Hot Water System.*

## *Energy-Efficient Boilers*

If you are building a new home or considering upgrading your existing boiler, be sure to study the many energy-efficient boilers now on the market. Most modern boilers can achieve combustion efficiencies in the 85% to 90% range. Some are as high as 92% to 95%.

If you can't afford a new boiler, you may want to consider an efficiency upgrade for your existing boiler. You can learn more about upgrades by reading DOE's Consumer's Guide on their Energy Efficiency and Renewable Energy website at www.eere.energy.gov.) Your first upgrade, however, should be insulating the pipe that distributes the heat.

In an effort to boost the efficiency of a gas-fired boiler, manufacturers now typically include an electronic ignition, eliminating the standing pilot light. Electronic ignitions produce a spark that ignites the burner when heat is required. High-efficiency boilers are also designed to burn fuels more efficiently and extract more heat from combustion gases. These features mean more heat per BTU of fuel burned.

Like energy-efficient furnaces, energy-efficient boilers come with sealed combustion chambers where gas or oil burns. Replacement air comes from the outdoors, not from room air. A small fan draws air into the combustion chamber and forces the exhaust gas out through the flue. (These are known as *induced-draft boilers.*) This design reduces the amount of dangerous — and potentially lethal — pollutants entering a building. Sealed combustion chambers also reduce the amount of air infiltration through leaks in the building envelope. This natural influx replaces air lost through the flue. (As room air enters the combustion chamber of a furnace or boiler and then escapes through the flue with combustion waste products, it depressurizes a building. Outside air is then drawn into the home to replace air lost through the flue.) Sealed combustion chambers reduce air leakage through the building envelope that occurs with natural draft furnaces.

To check out your options, call several local HVAC professionals for recommendations; obtain their literature, and study it carefully. I recommend buying high-efficiency models that have been on the market for a while, and hiring installers who have been in the business for a while, too. Check the Better Business Bureau for complaints. Hire an installer who has a proven track record — with high customer satisfaction — and who provides a good warranty on the furnace and his or her work. To learn more about your options, check out Energy Star's website (www.energystar.gov). Click on "Heating and Cooling" under products, then click on "Boilers" for a consumer product list. This site contains an enormous amount of information and can be confusing at first, so take your time. Study your options very carefully. Take notes. Learn as much as you can *before* you contact an installer, so you know what they are talking about.

## Shopping for a New Furnace or Boiler

When shopping for an energy-efficient furnace or boiler, remember that quality costs more. You may end up paying $500 to $1,000 more for a high-efficiency model. However, that investment could save you a substantial amount of money over the long-term. You can usually offset the higher initial cost within a few years through lower heat bills. After this period, the savings from a new furnace essentially becomes a source of income.

Remember, too, that the higher upfront investment also provides a hedge against rising fuel costs. To calculate your cost savings, you can use the Savings Calculator on the Energy Star website. You can also calculate the return on investment via the cost calculator on the website of the American Council for an Energy-Efficient Economy (www.aceee.org).

When considering a new energy-efficient furnace or boiler, be sure to look into state and local tax incentives or rebates from local utilities. Many incentives are available to homeowners that will help reduce the higher initial cost of a more efficient furnace. To check out incentives in your area, call your utility or the state energy office or ask local installers. You can also find information on the Database on State Incentives for Renewables and Efficiency at www.dsireusa.org. Installers should be able to provide information on incentives, but always check with other sources or even an accountant to be sure you've got it right.

When buying a new furnace or boiler, be sure that the system is properly sized to your home, especially if you have retrofitted your home for energy efficiency. HVAC contractors have a tendency to oversize furnaces and boilers. According to the DOE, heating systems in new American homes are often two or three times larger than required.

Unfortunately, oversized furnaces and boilers cycle on and off more frequently than properly sized systems. This, in turn, results in less-efficient operation. (It takes a little time for a furnace to reach maximum efficiency, so if it is cycling on and off, it runs at peak efficiency less often.) Start-ups also require more fuel. All of these factors result in more fuel consumption — which may offset efficiency gains. As a rule, a heating system should be no more than 25% larger than the calculated heat requirement.

Be sure to hire a highly reputable HVAC contractor — one who is capable of installing the equipment and doing a good job, but also of accurately calculating your heat demand. Be sure they take into account energy-efficiency measures you've incorporated, especially weatherization efforts and your solar system. After your new boiler or furnace is in place, be sure to follow maintenance recommendations and hire a professional to periodically inspect and maintain your system.

Call several reliable installers in your area to see what kinds of systems they offer, and what they recommend. You can find a contractor by searching the Yellow Pages or, better yet, ask a general contractor you trust for recommendations.

## What about a Heat Pump for Backup?

With growing awareness of the dangers of global warming and the continually rising cost of home heating fuels, many people are considering heat pumps — sometimes as a backup to solar heating options. As you shop, you will find two choices: ground-source and air-source heat pumps.

Heat pumps are devices that use refrigeration technology to extract heat from the environment. Extracted heat is then transferred into a building.

Fig. 10-5: *This geothermal system captures heat from the ground (geothermal heat) to heat a home. It can also be run in reverse to cool a home during the summer.*

Heat pumps can remove heat from the air surrounding a building — an air-source heat pump — or from the ground — a ground-source heat pump. Ground-source heat pumps are typically referred to as *geothermal systems* (Figure 10-5). Heat extracted from the air or the ground is distributed through a building either through ducts in a forced-air system, or through pipes in radiant floor or hot water baseboard systems.

In geothermal systems (i.e., ground-source heat pumps), pipes are buried in the ground either vertically or horizontally. A pump circulates a fluid through the pipes, which absorbs heat from the Earth. In many locations the ground below the frost line remains a constant 50° F, plus or minus a few degrees year round. (The warmer your average annual temperature, the warmer the temperature of the soil beneath your feet.)

DAN CHIRAS

Fig. 10-6: *Outdoor unit of an air-source heat pump that heats and cools my office and classroom at the Evergreen Institute.*

In air-source heat pumps, heat is drawn directly from the air outside a building — even on a cold winter day. The heat is then distributed through a building via ducts or pipes, as in other heating systems. Air-source heat pumps are able to extract heat from the environment on cold winter days because the refrigerant in the pump is colder than the air temperature. Heat moves from the slightly warmer outside air to the colder refrigerant. So, whenever the outside air is warmer than the refrigerant flowing through the pump, the latter will pick up heat.

The equipment in both geothermal and air-source heat pumps is powered by electricity. However, for every unit of electricity a geothermal system consumes during operation, it produces about four units of heat. (In contrast, an electric heater produces only one unit of heat for every unit of electricity it consumes.) Air-source heat pumps are slightly less efficient. For every unit of electricity they consume, they produce about three units of heat, give or take a little.

Because they operate on electricity, heat pumps require no combustion of fossil fuel. This eliminates the risk of combustion gases such as carbon monoxide seeping into a house and results in healthier indoor air. It also reduces the chances of a fire.

Heat pumps can also be used to cool buildings in the summer. When run in reverse, these systems extract heat from a building and deposit it either into the air, if you have an air-source heat pump, or the ground, in the case of a geothermal system. In addition, some of the heat from a heat pump can be used to heat water for in-home use. As a result, heat pumps serve three purposes: they heat buildings in the winter, they cool buildings in the summer, and they provide hot water year round.

Air-source and ground-source heat pumps can replace conventional furnaces and boilers in existing homes. They tie into existing heating distribution systems. If you are building a new home, consider installing a heat pump. Air-source heat pumps are typically installed in milder climates, although at least two manufacturers (Mitsubishi and Fujitsu) now produce units for use in pretty cold climates. Check with local installers to see if air-source heat pumps work in your area. Geothermal systems (ground-source heat pumps) are typically installed in colder climates.

Fig. 10-7: *This photo shows the indoor unit of an air source heat pump. They are mounted on walls in rooms you want to heat or cool*

DAN CHIRAS

So, should you install a heat pump to provide backup heat?

While heat pumps are more efficient and cleaner than conventional furnaces and boilers, geothermal systems cost more to install — about 25% to 100% more than a high-efficiency boiler or furnace (depending on location, local labor costs, and difficulty of the installation). Even so, they can save a substantial amount of money over the long-term. Air-source heat pumps are fairly economical to install and provide excellent savings, but may not be appropriate in extremely cold climates.

The higher cost of a geothermal system is due to the extensive excavation and/or drilling required to lay the heat-absorbing pipes in the ground. For new construction, this can be done before the finish grading is completed. Installation in an existing yard, especially a small one, is often more difficult and considerably more expensive. Pipes may need to be installed vertically to a depth of 200 feet. This requires a drill rig. Getting a drill rig into a backyard in an urban or suburban setting may be difficult or impossible. In such cases, a solar hot water system may be more economical, provided there's sufficient Sun-bathed roof area for installation of the collectors.

Be sure that your new system is properly sized to your home, that is, it takes into account all the measures you've incorporated to increase your home's energy efficiency, such as adding insulation and caulking air leaks in outside walls. As just mentioned, contractors tend to oversize heating systems, which results in lower operating efficiencies of heat pumps.

When shopping for a heat pump, look for an Energy Star-qualified model. Although a bit more expensive, the difference in initial cost will be offset by lower energy bills. To calculate your return on investment, visit www.aceee.org.

If you dramatically reduce heating requirements through air sealing and insulation and provide a significant amount of space heat via one of the solar technologies discussed in this book, a heat pump may not be worth the substantial investment. Why invest in a $20,000–$50,000 geothermal system to provide a couple hundred dollars worth of heat over the winter? You might be better off installing a smaller, energy-efficient furnace or boiler or an air-source heat pump.

If you still want a heat pump, research the various products and compare their performance. You can locate a heat pump installer in the Yellow Pages or by logging on to www.natex.org or www.acca.org. Ask for references and call them. Be sure to check with the Better Business Bureau to see if there have been any complaints against an installer before you sign a contract.

## Conclusion

We've now explored energy-efficiency measures, the major solar home heating technologies, and backup heating systems. As I've mentioned many times, the first step to heating your home with solar energy is efficiency: air sealing. The second is upgrading the insulation.

After sealing up and insulating your home, you have several options to pick from: passive solar, solar hot air, and solar hot water systems. You'll need to assess each option carefully.

Passive solar is ideal for new homes — provided they are oriented to true south. This design can provide free heat for life at little, if any, additional cost. They also help maintain cooler temperatures in the summer.

Passive solar retrofits are possible, provided there's plenty of south-facing wall that can be "opened up" by installing windows, or solarized by installing an attached sunspace. Don't forget to include passive solar if you are building an addition onto an existing home or business.

Solar hot air systems work well, too, and require less structural modification of an existing home than a passive solar retrofit. Solar hot air collectors can be mounted on the south-facing walls of a building or on racks on the ground and oriented to ensure maximum solar gain.

Solar hot water systems (thermal systems) are ideal in applications in which the south-facing walls are shaded and therefore not amenable to passive solar or solar hot air systems. Of course, you need to have room on a sunny roof to install solar hot water collectors and room in your house to install the large solar storage tank. Solar collectors can also be installed on the ground next to a building and oriented for maximum solar gain.

All these technologies offer a very decent return on investment on their own, but they may also qualify for financial incentives, including tax credits and rebates. Financial incentives such as these lower the initial cost and make your investment in solar even more profitable.

Combined with energy-efficiency measures and an energy-efficient backup heating system, solar home heating systems can help save you money — a substantial amount of money — and dramatically reduce your environmental footprint, creating a greener world for the benefit of present and future generations and the millions of species that share planet Earth with us.

# Resource Guide

## Chapter 1: Introduction

Aitken, Donald W. "The Renewable Energy Transition: Can it Really Happen?" *Solar Today* 19(1), 16–19, 2005. A valuable look at a very important question.

Aitken, Donald W. "Germany Launches Its Transition," *Solar Today* 19(2), 26–19, 2005. A fascinating look at Germany's plans to achieve 100% of its energy from renewable sources.

Asmus, Peter. "Power Solutions in Our Own Backyards," *Solar Today* 15 (1), 2001. An interesting look at dispersed energy production.

Bihn, Dan. "Japan Takes the Lead," *Solar Today* 19(1), 20–23, 2005. A look at Japan's aggressive pursuit of renewable energy.

Chiras, Dan and J. Richter. "Money Matters: Does an RE System Make Economic Sense?" *Home Power* 131, 82–85, 2009. Discusses several ways to analyze the cost-effectiveness of a renewable energy system.

Campbell, C. J. "The Oil Peak: A Turning Point," *Solar Today* 15(4), 40–43, 2001. A brief analysis of the future of global oil.

Darley, Julian. *High Noon for Natural Gas: The New Energy Crisis.* Chelsea Green, 2004. A look at natural gas supplies and the impacts of the imminent peak in global natural gas production.

Hartmann, Thom. *The Last Hours of Ancient Sunlight: Warming*

*Up to Personal and Global Transformation*. New York: Three Rivers Press, 1999. A philosophic look at the decline in oil.

Heinberg, Richard. *The Party's Over: Oil, War and the Fate of Industrial Societies*. New Society Publishers, 2003. A grim look at the prospects of human society as global oil production peaks.

Heinberg, Richard. *Power Down: Options and Actions for a Post-Carbon World*. New Society Publishers, 2004. A frank discussion of our options as global oil production peaks.

Home Power staff. "Clearing the Air: Home Power Dispels the Top RE Myths," *Home Power* 100, 32–38, 2004. Superb piece for those who want to learn more about the many false assumptions and beliefs about renewable energy.

Prugh, Tom, Christopher Flavin, and Janet L. Sawin. "Changing the Oil Economy" In *State of the World 2005*. Starke, L. (ed.). W. W. Norton, 2005. Examines oil production declines from the standpoint of national security.

Sawin, Janet L. "Charting a New Energy Future," In *State of the World 2003*. Starke, L. (ed.). Norton, 2003. An overview of renewable energy potential and what other nations are doing to create a sustainable energy future.

Sawin, Janet L. "Mainstreaming Renewable Energy in the 21st Century," Worldwatch Paper 169, Washington, DC: Worldwatch Institute, 2004. Overview of the prospects of renewable energy.

Sawin, Janet L. "Making Better Energy Choices," In *State of the World 2004*. Starke, L. (ed.). Norton, 2004. An overview of what people are doing worldwide to switch to a renewable energy strategy.

Sindelar, Allan and Phil Campbell-Graves. "How to Finance Your Renewable Energy Home," *Home Power* 103, 94–99, 2004. Very useful article.

Udall, Randy and Steve Andrews. "Methane Madness: A Natural Gas Primer," *Solar Today* 15(4), 36–39, 2001. A frightening look at natural gas production.

Udall, Randy and Steve Andrews. "When Will the Joyride End?," *Home Power* 81, 43–48, 2001. An insightful look at oil production and consumption with emphasis on peak oil production.

## CHAPTERS 2, 3, AND 4: ENERGY, ENERGY EFFICIENCY, AND ENERGY CONSERVATION

Aman, Jennifer T. and Alex Wilson. "Home Heating Basics," *Home Power* 123, 50–55. A great little article on home heating options.

Bailes, Allison A. "How Efficient is Your House?" *Home Power* 106, 74–78, 2005. Great article on home energy ratings and blower door tests.

Chiras, Dan. *Green Home Improvement: 65 Projects that Will Cut Utility Bills, Protect Your Health, and Help the Environment*. R. S. Means: Kingston, MA, 2008. A great set of projects you can carry out around your home to improve its energy efficiency.

Chiras, Dan. "All About Insulation," *Mother Earth News* 195, 62–70, 2003. A detailed look at insulation options.

Chiras, Dan. "Make Your Home Energy Efficient," *Mother Earth News*, October/November, 84–92, 2008. My advice on energy retrofitting a home.

Chiras, Daniel D. "The Energy-Efficient Home," *Solar Today* 18(5), 22–27, 2004. A primer on energy-efficient home design.

Carmody, John, Stephen Selkowitz, and Lisa Heschong. *Residential Windows: A Guide to New Technologies and Energy Performance*. Norton, 1996. Important resource for energy-efficient home designers.

Crume, Richard. "Home Energy Upgrades That Pay," *Solar Today* 23(2): 28–33, 2010. Great discussion of the economics of energy retrofitting.

Fine Homebuilding. *The Best of Fine Homebuilding: Energy-Efficient Building*. Taunton Press, 1999. A collection of detailed articles on a wide assortment of topics related to energy efficiency including insulation, energy-saving details, windows, and heating systems.

Hurst-Wajszczuk, Joe. "Save Energy and Money — Now!" *Mother Earth News* 189, 24–33, 2001. Ideas on ways to save energy in your home.

Kerr, Andy. "Doing Well While Doing Good: Conservation of Energy as a Rational Financial Investment," *Home Power* 86, 96–103, 2002. A look at the economics of household energy savings.

Johnston, David and Kim Master. *Green Remodeling: Changing the World One Room at a Time*. New Society Publishers, 2004. Superb coverage of many ideas on ways to boost energy efficiency in existing homes.

Lstiburek, Joe. *EEBA Builder's Guide to Cold Climates*. Energy Efficient Building Association, 1999. Superb resource for advice on building in cold climates.

______. *EEBA Builder's Guide to Mixed Humid Climates*. Energy Efficient Building Association, 1999. Superb resource for advice on this climate.

______. *EEBA Builder's Guide to Hot-Arid Climates*. EEBA, 1999. Superb resource for advice on building in hot arid climates.

National Association of Home Builders Research Center. *Design Guide for Frost-Protected Shallow Foundations*. NAHB Research Center, 1996. Also available online.

Pahl, Greg. *Home Heating Basics*. Chelsea Green, 2003. A detailed overview of natural home heating options.

Reysa, Gary. "8 Easy Projects for Instant Energy Savings," *Mother Earth News*, February/March, 54–60, 2008. Great advice for all homeowners.

Scheckel, Paul. *The Home Energy Diet*. New Society Publishers, 2005. A great guide for energy conservation in homes.

Scheckel, Paul. "Efficiency Details for a Clean Energy Change," *Home Power* 121, 40–45, 2007. Great article, well worth reading.

Yewdall, Zeke. "Alternatives that Don't Cost an Arm and a Leg," *Home Power* 101, 96–100, 2004. A useful article on energy efficiency and other ways to acquire electricity from renewable sources if your site is not ideally suited or your pocketbook prohibits a big investment.

Yost, Harry. *Home Insulation: Do It Yourself and Save as Much as 40%*. Storey Communications, 1991. Extremely useful book to read for anyone building a new house.

## CHAPTER 5: UNDERSTANDING SOLAR ENERGY

Chiras, Dan. *Power from the Sun*. New Society Publishers, 2009. Contains a detailed chapter on the Sun and solar energy.

## CHAPTERS 6 AND 7: PASSIVE SOLAR HEATING

Chiras, Daniel D. "Build a Solar Home and Let the Sunshine in," *Mother Earth News* 193, 74–81, 2002. A survey of passive solar design principles and a case study showing the economics of passive solar heating.

_____. *The Solar House: Passive Heating and Cooling*. Chelsea Green, 2002. A detailed, readable guide for designing and building homes for passive solar heating and cooling.

_____. "Sun-Wise Design: Avoid Passive Solar Design Blunders," *Home Power* 105, 38–44, 2005. Important look at the most costly and most common mistakes in passive solar design.

_____. "Passive Solar Retrofit," *Home Power* 138, 106–111, 2011. Explains ways you can retrofit an existing building with passive solar.

Haggard, Ken. "Basics of Passive Solar Design," *Solar Today* 22(3): 6A–9A, 2008. A good discussion of passive solar.

Haggard, Ken and David Bainbridge. "Passive Solar Means Efficiency and Use of On-Site Energy," *Solar Today* 23(8), 32–33, 2010.

Johnston, David and Kim Master. *Green Remodeling: Changing the World One Room at a Time*. New Society Publishers, 2004. Many ideas that boost energy efficiency in existing homes and increase reliance on renewable energy, including passive solar.

Kachadorian, James. *The Passive Solar House*. Chelsea Green, 1997. Presents a lot of good information on passive solar heating and an interesting design that has been fairly successful in cold climates.

Kipnis, Nathan. "You Can Build this Energy-Efficient Solar Home," *Mother Earth News*, August/September, 58–63, 2009. An article well worth reading. Contains lots of good ideas.

Little, Amanda Griscom. "Super Solar Homes Everyone Can Afford," *Mother Earth News* 207, 34–41, 2005. Those interested in building a new home should check out this article.

Miller, Burke. *Solar Energy: Today's Technologies for a Sustainable Future*. American Solar Energy Society, 1997. Extremely valuable resource; contains numerous case studies showing how passive solar heating can be used in different climates, even in some solar-deprived places.

Olson, Ken and Joe Schwartz. "Home Sweet Solar Home: A Passive Solar Design Primer," *Home Power* 90, 86–94, 2002. Superb introduction to passive solar design principles.

Riversong, Robert." Designing a Passive Solar Slab," *Home Power* 136, 60–66, 2010. A detailed discussion of ways to insulate slabs that serve as thermal mass for passive solar homes.

Reynolds, Michael. *Comfort in Any Climate*. Solar Survival Press, 1990. A brief, but informative, treatise on passive heating and cooling.

Reysa, Gary. "Solar Heating Plan for Any Home," *Mother Earth News*, Special Issue (spring), 60-65, 2008.

Sklar, Scott and Kenneth Sheinkopf. *Consumer Guide to Solar Energy: More Ways to Reduce Your Energy Bills and Save the Environment*. Bonus Books, 1995. Delightful introduction to many different solar applications, including passive solar heating.

Sustainable Buildings Industry Council. *Designing Low-Energy Buildings: Passive Solar Strategies and Energy-10 software*. SBIC, 1996. A superb resource! This book of design guidelines and the *Energy-10* software that comes with it enables builders to analyze the energy and cost savings in building designs. Helps permit region-specific design.

## Chapter 8: Solar Hot Air

Chiras, Dan. *The Homeowner's Guide to Renewable Energy.* New Society Publishers, 2006. A good overview of numerous solar technologies, including solar hot air systems.

Mehalic, Brian. "Drainback Solar Hot Water Systems, *Home Power,* 138, 78–85. 2011. Superb article on drainback systems. Great illustrations.

Marken, Chuck. "Solar Hot Air System Design," *Home Power,* Dec. 2004/January 2005, 68 -72.

_____. "Solar Hot Air Systems: Part II," *Home Power,* February/March, 2005, 88–94.

_____. "Overcoming Overheating," *Home Power,* 142, 88–91, 2011. Excellent set of ideas on what to do with excess heat generated by a solar hot water system.

Reysa, Gary. "Build a Solar Heater for $350," *Home Power,* October/November. 2005, 30–35.

Wilson, A., J. Thorne, and J. Morril. *Consumer Guide to Home Energy Savings,* American Council for an Energy-Efficient Economy, 2003.

## Chapter 9: Solar Hot Water Heating

Butler, Barry. "Solar Wand: Hot Water Assist for Cold Climates," *Home Power* 104, 86–91, 2005. Illustrates and describes a quick retrofit for existing water tanks.

Galloway, Terry. *Solar House: A Guide for the Solar Designer.* Architectural Press, 2004. A technical guide on a wide range of solar designs, including active solar.

Hyatt, Rod. "Hydronic Heating on Renewable Energy," *Home Power* 79, 36–42, 2000. Provides a lot of practical advice on building your own radiant floor heating system and powering it with photovoltaic panels.

Lane, Tom and Ken Olson. "Solar Hot Water for Cold Climates: Part II — Drainback Systems," *Home Power* 86, 62–70, 2002. Detailed look at drainback systems. (Part I of this series is Olson's article on closed-loop antifreeze systems, listed below. Part 3, written by Chuck Marken and Olson, is also listed below.)

Marsden, Guy. "Solar Heat for My Main Workshop," *Home Power* 89, 34–43, 2002. A case study worth reading by those interested in using solar hot water to provide space heat.

Marken, Chuck. "Backup Electric Water Heaters," *Home Power* 137, 66–70, 2010. Great discussion of energy-efficient electric water heaters.

______. "Heat Exchangers for Solar Water Heating," *Home Power* 92, 68–76, 2003. Great overview of heat exchangers, including the types that are available for use in solar hot water systems.

______. "Solar Collectors: Behind the Glass," *Home Power* 133, 70–76, 2009. Interesting coverage of the anatomy of modern solar hot water collectors.

______. "Solar Hot Water," *Home Power*, 141, 88–91, 2011. Excellent overview of solar hot water systems. Very well illustrated.

Marken, Chuck and Ken Olson. "Installation Basics for Solar Domestic Water Heating Systems," *Home Power* 94, 50–57, 2003. The first in a series of three articles for those who would like to install their own solar hot water systems.

Marken, Chuck and Ken Olson. "SDHW Installation Basics. Part 3: Drainback System," *Home Power* 97, 48–54, 2003. Excellent reference for installers and do-it-yourselfers.

Mehalic, Brian. "Solar Hot Water Storage," *Home Power* 131, 70–78, 2009. A detailed look at solar hot water storage, especially heat exchangers.

______. "Flat-Plate and Evacuated-Tube Solar Thermal Collectors," *Home Power* 132, 40–46, 2009.

Olson, Ken. "Solar Hot Water: A Primer," *Home Power* 84, 44–52, 2001. Excellent overview of solar hot water systems and your options.

Olson, Ken. "Solar Hot Water for Cold Climates: Closed-Loop Antifreeze System Components," *Home Power* 85, 40–48, 2001. For those interested in installing a solar hot water system in a climate where wintertime freezing is a regular occurrence.

Owens, Bob. "Florida Batch Water Heater," *Home Power* 93, 66–70, 2003. For those interested in installing a solar batch heater.

Patterson, John and Suzanne Olsen. "Single Tank Solar Water Systems," *Home Power* 124, 42–46, 2008. This article presents another option for homes that don't have room for a solar storage tank.

Perlin, John. "Solar Hot Water History," *Home Power* 100, 64–67, 2004. A great overview of the history of solar hot water.

Pine, Nick. "Solar Heat in Snow Country," *Solar Today* 17 (1), 36–37, 2003. An inspiring story about active solar heating in a US Customs border station in Vermont.

Ramlow, Bob and Benjamin Nusz. *Solar Water Heating: A Comprehensive Guide to Solar Water and Space Heating Systems*. New Society Publishers, 2006.

Reysa, Gary. "Solar Heating Plan for Any Home," *Mother Earth News* Special Issue (spring), 60–65, 2008. You'll find this innovative idea very interesting.

Sklar, Scott and Kenneth Sheinkopf. *Consumer Guide to Solar Energy: More Ways to Reduce Your Energy Bills and Save the Environment*. Bonus Books, 1995. Delightful introduction to many different solar applications, including solar hot water.

Simpson, Walter. "Adventure in Solar Living," *Solar Today* 17(5), 29–32, 2003. An inspiring tale of a passive solar/solar hot water-heated home in Buffalo, New York.

Sindelar, Allan and Phil Campbell-Graves. "How to Finance Your Renewable Energy Home," *Home Power* 103, 94–99, 2004. Very useful article.

Sklar, Scott. "Selecting a Solar Heating System," *Solar Today* 18(5), 42–45, 2004. A good look at the economics of solar hot water systems.

## CHAPTER 10: ENERGY-EFFICIENT BACKUP HEAT

Chiras, Dan. *Green Home Improvement: 65 Projects that Will Cut Utility Bills, Protect Your Health, and Help the Environment.* R. S. Means, Kingston, MA, 2008. Contains tons of information on energy efficiency and heating systems.

Malin, Nadav and Alex Wilson. "Ground-Source Heat Pumps: Are They Green?" *Environmental Building News* 9(1), 16-22, 2000. Detailed overview on ground-source heat pumps.

National Renewable Energy Lab. "Geothermal Heat Pumps," published online at www.eren.doe.gov/erec/factsheets/geo_heatpumps.html. Great overview of ground-source heat pumps.

Persons, Jeff. "The Big Dig," *Mother Earth News* 185, 52–54, 102. A brief introduction to ground-source heat pumps.

# Index

## About the Author

DAN CHIRAS is an internationally acclaimed author who has published 30 books, including *The Homeowner's Guide to Renewable Energy, Power From the Sun, Green Home Improvement,* and *The Solar House.* He is a certified wind site assessor and has installed numerous residential wind and solar electric systems. Dan is director of The Evergreen Institute's Center for Renewable Energy and Green Building (www.ever greeninstitute.org) in east-central Missouri where he teaches workshops on small wind energy systems, solar electricity, passive solar design and green building. Dan also has an active consulting business, Sustainable Systems Design (www.danchiras.com) and has consulted on numerous projects in North America and Central America in the past ten years. Dan lives in a net zero energy home powered entirely by solar and wind.

*Dan Chiras*

If you have enjoyed *Solar Home Heating Basics*, you might also enjoy other

# BOOKS TO BUILD A NEW SOCIETY

Our books provide positive solutions for people who want to make a difference. We specialize in:

**Sustainable Living • Green Building • Peak Oil • Renewable Energy • Environment & Economy Natural Building & Appropriate Technology Progressive Leadership • Resistance and Community Educational & Parenting Resources**

## New Society Publish...



E... T

New S... d paper r... ne free, and old growth free.

For every 5,000 books printed, New Society saves the following resources:[1]

| | |
|---|---|
| 15 | Trees |
| 1,357 | Pounds of Solid Waste |
| 1,494 | Gallons of Water |
| 1,948 | Kilowatt Hours of Electricity |
| 2,468 | Pounds of Greenhouse Gases |
| 11 | Pounds of HAPs, VOCs, and AOX Combined |
| 4 | Cubic Yards of Landfill Space |

[1]Environmental benefits are calculated based on research done by the Environmental Defense Fund and other members of the Paper Task Force who study the environmental impacts of the paper industry.

*For a full list of NSP's titles, please call* 1-800-567-6772 *or check out our website* at:

**www.newsociety.com**